Sustainable and Advanced Applications of Blockchain in Smart Computational Technologies

This book concentrates on the sustainable applications of the blockchain technology across multiple latest computational knowledge domains. It covers the feasible and practical collaboration of blockchain technology with the latest Sustainable Smart Computing Technologies. It will target the vast applications of blockchain in the field of Internet of Things, Artificial Intelligence, and Cybersecurity. The book effectively provides satisfactory information about the essentials of blockchain and IoT to a typical pursuer alongside encouraging an examination researcher to distinguish some modern issue regions that rise up out of the intermingling of the two advancements. Besides, the creators talk about pertinent application zones, for example, smart city and e-social insurance, along the course of the book.

- Covers the recent advancements in blockchain technology
- Discusses the applications of Blockchain technology for real-life problems
- Addresses the challenges related to implementation of Blockchain technology
- Includes case studies
- Includes the latest trends and area of research in blockchain technology

This book is primarily aimed at graduates, researchers, and professions working in the field of blockchain technology.

Sustainable and Advanced Applications of Blockchain in Smart Computational Technologies

Edited by
Keshav Kaushik
Shubham Tayal
Susheela Dahiya
Ayodeji Olalekan Salau

CRC Press
Taylor & Francis Group
Boca Raton London New York

CRC Press is an imprint of the
Taylor & Francis Group, an **informa** business

A CHAPMAN & HALL BOOK

First edition published 2023
by CRC Press
6000 Broken Sound Parkway NW, Suite 300, Boca Raton, FL 33487-2742

and by CRC Press
4 Park Square, Milton Park, Abingdon, Oxon, OX14 4RN

CRC Press is an imprint of Taylor & Francis Group, LLC

Library of Congress Cataloging-in-Publication Data
Names: Kaushik, Keshav, editor. | Tayal, Shubham, editor. |
Dahiya, Susheela, editor. | Salau, Ayodeji Olalekan, editor.
Title: Sustainable and advanced applications of blockchain in smart computational technologies /
edited by Keshav Kaushik, Shubham Tayal, Susheela Dahiya, Ayodeji Olalekan Salau.
Description: First edition. | Boca Raton : Chapman & Hall/CRC Press, 2023. |
Includes bibliographical references. |
Summary: "This book concentrates on the sustainable applications of the Blockchain Technology across multiple latest computational knowledge domains. It covers the feasible and practical collaboration of Blockchain Technology with latest Sustainable Smart Computing Technologies. It will target the vast applications of Blockchain in the field of Internet of Things, Artificial Intelligence, and Cybersecurity. The book effectively provides satisfactory information about the essentials of Blockchain and IoT to a typical pursuer alongside encouraging an examination researcher to distinguish some modern issue regions that rise up out of the intermingling of the two advancements. Besides, the creators talk about pertinent application zones, for example, smart city, e-social insurance, and so forth along the course of the book"—Provided by publisher.
Identifiers: LCCN 2022015570 (print) | LCCN 2022015571 (ebook) | ISBN 9781032044217 (hardback) |
ISBN 9781032044873 (paperback) | ISBN 9781003193425 (ebook)
Subjects: LCSH: Blockchains (Databases)
Classification: LCC QA76.9.B56 S87 2023 (print) | LCC QA76.9.B56 (ebook) |
DDC 005.74—dc23/eng/20220714
LC record available at https://lccn.loc.gov/2022015570
LC ebook record available at https://lccn.loc.gov/2022015571

ISBN: 9781032044217 (hbk)
ISBN: 9781032044873 (pbk)
ISBN: 9781003193425 (ebk)

DOI: 10.1201/9781003193425

Typeset in Palatino
by codeMantra

Contents

Preface

Blockchain technology creates a network that is decentralized, distributed, and secure. In blockchain technology, each cluster is linked in a decentralized peer-to-peer network, where each transaction is immediately recorded with date stamps and operations are distributed without any outside influence. Farming, medical, security, and banking are all sectors where the blockchain method might be beneficial. Through cryptographic hashing, the data provided in blocks is further connected and safeguarded in chains with digital signatures. Because each block is linked to the previous block, hackers will be unable to hijack transactions by injecting harmful data into the system. Digital signatures, validation, smart contracts, decentralization, safe collaboration, and immutable understandable artificial intelligence have all been handled by combining the blockchain method with machine learning for Internet of Things frameworks. With the development of smart IoT devices and their interconnectivity, massive amounts of data are now being generated in a consolidated format. As a result, technological advancements often produce difficulties such as space, security, and privacy.

Blockchain technology allows for the distributed, reliable, safe, and decentralized exchange of ledger data. The blockchain method's decentralized storage is used to store enormous amounts of data that are linked to the current block and previous blocks through smart contracts. For both small and big businesses, blockchain may be used to increase security, privacy, and data openness. Blockchain is a well-known technology that has the potential to improve the industrial and supply chain environments. Various disciplines now offer intriguing insights on blockchain's benefits.

Sustainable and Advanced Applications of Blockchain in Smart Computational Technologies is intended to describe and explain the advanced applications of blockchain technology in various Smart Computational Technologies, with the intent of making the reader aware of the constraints and considerations applied in blockchain technology. Upon reading this book, the reader should have a proper overview of the field of blockchain technology, starting them on the journey of becoming a Blockchain Developer and expert.

This book is a decent assortment of best-in-class approaches used in the fields of Blockchain, Cybersecurity, Digital Forensics, Artificial Intelligence, and Internet of Things. It will be valuable for the new scientists and specialists working in the field to rapidly know the best performing strategies. It will help them in realizing their ideas, analyze various approaches, and deliver the best one to the society. This book will be helpful as textbook/reference book for undergraduate and postgraduate students as a number of reputed universities have blockchain technology as a part of their curriculum these days.

Editors

Mr. Keshav Kaushik is an Assistant Professor in the School of Computer Science at the University of Petroleum and Energy Studies, Dehradun, India. He is an educator with over seven years of teaching and research experience in Cybersecurity, Digital Forensics, the Internet of Things, and Blockchain Technology. Mr. Kaushik received his B.Tech. degree in Computer Science and Engineering from the University Institute of Engineering and Technology, Maharshi Dayanand University, Rohtak. In addition, he also has an M.Tech. degree in Information Technology from YMCA University of Science and Technology, Faridabad, Haryana. He has published 20+ research papers in international journals and has presented at reputed international conferences. He is a Certified Ethical Hacker (CEH) v11, CQI and IRCA-Certified ISO/IEC 27001:2013 Lead Auditor, and Quick Heal Academy-Certified Cyber Security Professional (QCSP). He acted as a keynote speaker and delivered 50+ professional talks on various national and international platforms. He has edited more than seven books with reputed publishers.

Dr. Shubham Tayal is an Assistant Professor in the Department of Electronics and Communication Engineering at SR University, Warangal, India. He has more than six years of academic/research experience teaching at the UG and PG levels. He has received his Ph.D. in Microelectronics & VLSI Design from National Institute of Technology, Kurukshetra; M.Tech. (VLSI Design) from YMCA University of Science and Technology, Faridabad; and B.Tech. (Electronics and Communication Engineering) from MDU, Rohtak. He has qualified GATE (2011, 2013, 2014) and UGC-NET (2017). He has published more than 25 research papers in various international journals and conferences of repute, and many papers are under review. He is on the editorial and reviewer panel of many SCI/SCOPUS-indexed international journals and conferences. Currently, he is the editor of four books from CRC Press (Taylor & Francis Group, USA). He acted as keynote speaker and delivered professional talks on various forums. He is a member of various professional bodies like IEEE, IRED, etc. He is on the advisory panel of many international conferences. He is a recipient of Green ThinkerZ International Distinguished Young Researcher Award 2020. His research interests include simulation and modeling of multi-gate semiconductor devices, device-circuit co-design in digital/analog domain, machine learning, and IoT.

Dr. Susheela Dahiya is currently working as an Assistant Professor (Selection Grade) in the School of Computer Science at the University of Petroleum & Energy Studies, Dehradun, Uttarakhand, India. She received her M.Tech. (Computer Science & Engineering) in 2008 and Ph.D. in 2015 from the Indian Institute of Technology Roorkee. She has also qualified GATE and NET in Computer Science. She has more than nine years of academic/research/ industry experience. Her research interests include image & video processing, IoT, cyber security, cloud computing, and deep learning. She has authored several research papers in renowned conferences, Scopus & SCI journals.

Dr. Ayodeji Olalekan Salau received his B.E. in Electrical/Computer Engineering from the Federal University of Technology, Minna, Nigeria. He received M.Sc. and Ph.D. degrees from the Obafemi Awolowo University, Ile-Ife, Nigeria. His research interests include computer vision, image processing, signal processing, machine learning, power sys-tems technology, and nuclear engineering. Dr. Salau serves as a reviewer for several repu-table international journals. His research has been published in many reputable interna-tional conferences, book chapters, and major international journals. He is a registered engineer with the Council for the Regulation of Engineering in Nigeria (COREN), a member of the International Association of Engineers (IAENG), and a recipient of the Quarterly Franklin Membership with ID number CR32878 given by the Editorial Board of London Journals Press in 2020 for top-quality research output. More recently, Dr. Salau's paper was awarded the best paper of the year 2019 in Cogent Engineering. Currently, Dr. Salau works at Afe Babalola University in the Department of Electrical/ Electronics and Computer Engineering. In addition, he is the recipient of the International Research Award on New Science Inventions (NESIN) under the category of "Best Researcher Award" given by ScienceFather with ID number 9249, 2020.

Contributors

M. A. Adeagbo
Department of Mathematics and
    Computer Sciences
First Technical University
Ibadan, Nigeria

S. A. Akinseinde
Software Development Unit
The Amateur Polymath
Lagos, Nigeria

J. E. T. Akinsola
Department of Mathematics and
    Computer Sciences
First Technical University
Ibadan, Nigeria

Tanmay Bhowmik
School of Computer Science
    Engineering and Technology
Bennett University
Greater Noida, India

Kuldeep Chaurasia
School of Computer Science
    Engineering and Technology
Bennett University
Greater Noida, India

Madu Afam Daniel
Lion Science Park
University of Nigeria
Nsukka, Nigeria

Bamidele Sunday Fakinle
Department of Chemical
    Engineering
Landmark University
Omu-Aran, Nigeria

Olayomi Abiodun Falowo
Department of Chemical
    Engineering
Landmark University
Omu-Aran, Nigeria

Smriti Gaba
Reliance Jio Infocomm Limited
Navi Mumbai, India

Nithin Kamineni
School of Computer Science
    Engineering and Technology
Bennett University
Greater Noida, India

Avita Katal
School of Computer Science
University of Petroleum & Energy
    Studies (UPES)
Dehradun, India

T. Mahalakshmi
School of Computer Science
    Engineering and Technology
Bennett University
Greater Noida, India

Veera Nitish Mattaparthi
School of Computer Science
    Engineering and Technology
Bennett University
Greater Noida, India

Ambika Nagaraj
Department of Computer Science
    and Applications
St. Francis College
Bangalore, India

Oludare Johnson Odejobi
Department of Chemical
    Engineering
Obafemi Awolowo University
Ile -Ife, Nigeria

Ebenezer Leke Odekanle
Department of Chemical and
    Mineral Resources Engineering
First Technical University
Ibadan, Nigeria

Atanda Aminat Oluchi
Department of Computer Science
University of Nigeria
Nsukka, Nigeria

F. O. Onipede
Department of Mathematics and
    Computer Sciences
First Technical University
Ibadan, Nigeria

Vamsi Pachamatla
School of Computer Science
    Engineering and Technology
Bennett University
Greater Noida, India

A. Mona Reddy
School of Computer Science
    Engineering and Technology
Bennett University
Greater Noida, India

Jeevesh Sharma
Department of Commerce
Manipal University
Jaipur, India

Rishi Raj Singh
School of Computer Science
University of Petroleum & Energy
    Studies (UPES)
Dehradun, India

Andrea Milagros Carrasco Suyo
Academic program of Business
    Administration
University of Piura
Lima, Peru

Manish Thakral
School of Computer Science
University of Petroleum & Energy
    Studies (UPES)
Dehradun, India

Samarth Vashisht
Synopsys Inc.
Karnataka, India

Suhasini Verma
Department of Business
    Administration
Manipal University
Jaipur, India

A. A. Yusuf
Department of Information and
    Communication Technology
Federal University of Petroleum
    Resources
Effurun, Nigeria

1

Blockchain Platforms, Architectures, and Consensus Algorithm

Atanda Aminat Oluchi

Department of Computer Science, University of Nigeria

Madu Afam Daniel

Lion Science Park, University of Nigeria

CONTENTS

1.1 Introduction

Blockchain technology is a new revolutionary way to address issues like transparency, trust, data security, and access control (Lopes & Pereira, 2019). According to Wikipedia, blockchain is a continuously expanding set of records (database) called blocks that are connected together. Satoshi Nakamoto built the first blockchain-based system commonly recognized as a cryptocurrency (Bitcoin) in July 2008.

DOI: 10.1201/9781003193425-1

Blockchain is an encrypted ledger stored in a publicly distributed database. The ledger is encrypted to keep the identities of network (node) participants fully hidden. A ledger consists of a world state that holds the current value of all things and a blockchain that records the history of all transactions that led to the present world state; the ledger system is totally electronic and very secure (Attaran & Gunasekaran, 2019). Each member of the blockchain network (nodes) has a full copy of all records and ensures that all records and procedures are in order, resulting in data validity and security (Yaga et al., 2018).

When a block is created, all of the network's nodes must agree on the block's validity before it can be added to the network to join the chain of blocks already existing. This is what a consensus policy/protocol is all about. Proof of work, proof of stake (PoS), Byzantine fault-tolerant replication, and other consensus techniques are used in the blockchain. Every consensus algorithm guarantees that the data on the ledger is correct and ensures that all nodes in the network have the same data on the ledger, preventing bad actors from tampering with it.

Apart from Bitcoin, there are other blockchain technologies that are already in use. Understanding the architecture utilized to create these existing platforms can aid in the development of a more robust blockchain platform capable of solving additional specific challenges.

1.2 Blockchain Architecture

The word blockchain is made up of two words: block and chain.

In blockchain, block represents records of data, while chain represents how these records of data are linked together.

A *block* represents records of data of transactions that has happened. That is, once transactions happen, records of the transaction are stored in a block. The block is identified by a hashcode. Once the transaction is verified, then the block becomes a permanent part of the blockchain. This process continues to happen to grow a chain of blocks.

New blocks are always added to the end of the chain, i.e., a new block is always going to be block number $n+1$, where n is the number of blocks already existing. Blocks in a blockchain are cryptographically linked with each other through hashes.

In a chain of blocks, the first block is called the *Genesis Block*. The Genesis Block can't point to a previous hash but other blocks point to a previous hash rather it points to its previous hash as 0's depending on the byte of the hash. Therefore, Genesis Block is the only block without a previous hash but other blocks that follow the Genesis Block always point to a previous hash which makes the blockchain secured.

As shown in Figure 1.1, a block majorly is comprised of:

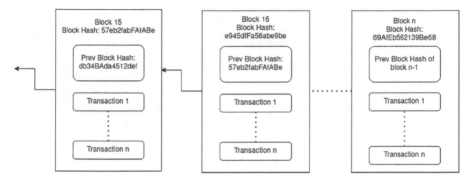

FIGURE 1.1
Chain of blocks. (Based on Yaga et al. (2018).)

a. Block numbers

b. **Present Block Hashcode:** This is the identifying factor of a block.

c. **Previous Block Hashcode:** This is the hashcode of the previous block before the present block. This is part of blockchain security measures by making present blocks have the hashcode of the previous block. This also creates the chain of blocks in other words a link of these blocks is created. Since the ledger is distributed if on one node end block B has a previous hashcode of "11000…" and another node has the same ledger but block B has a previous hashcode that is different, this automatically throws an error and the block with the tampered detail will not be verified or added to the block.

d. Transactions that have happened.

1.2.1 Ledger in Blockchain

A ledger consists of a world state, which contains the current value of all objects, and a blockchain that records the history of all transactions that resulted in the current world state. As shown in Figure 1.2, a ledger consists of two data structures:

a. Blockchain, which is a linked list of blocks/records of all transaction history that resulted in the world state. Each block describes a set of transactions that can't be tampered with.

b. World state, which is an ordinary database usually NoSQL database that contains key/value stores and stores the combined outputs of all transactions. World state is the current moment in time of what the network looks like.

Blockchain is built on various technologies such as distributed computing and decentralization. These technologies are described in detail in the

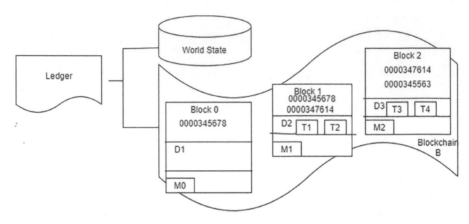

FIGURE 1.2
Ledger. (Yaga et al. (2018).)

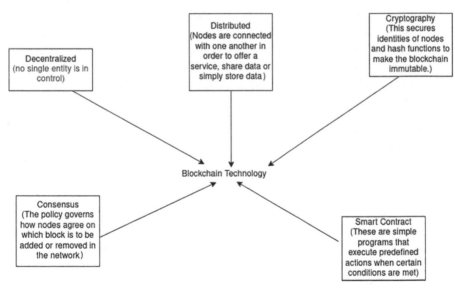

FIGURE 1.3
Concepts that make blockchain secured.

following sections to show how each of them has played a role in making blockchain a secured technology (Figure 1.3).

1.2.2 Distributed System

A distributed system is a collection of independent, geographically dispersed, networked computing units (nodes) that seem as a single coherent system to its users (Steen & Tanenbaum, 2016). There are two types of distributed system architecture:

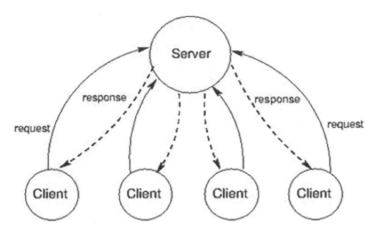

FIGURE 1.4
Client-server architecture. (Steen & Tanenbaum et al. (2016).)

- **Client-Server Architecture:** The Server and Clients are the two fundamental entities in this architecture, as shown in Figure 1.4, and they communicate with each other using the Request-Response technique.

  a. **Server:** This is the entity of the client-server architecture in charge of providing services to the client. The server serves as the main centralized entity responsible for fulfilling all requests from all clients in the same network who are connected to the server. The server performs tasks such as data processing, storage, and application deployment.

  b. **Clients:** This is a component of the client-server architecture that communicates with the server to perform a task. They are typically connected to an Internet server.

  The most significant downside of this architecture is that the entire system is reliant on a single point (i.e., server). If the server goes down, the entire system goes down with it.

- **Peer-to-Peer Architecture:** Figure 1.5 shows the architecture of a distributed system's peer-to-peer (P2P) architecture. This architecture functions similarly to a network of networked computer systems that can exchange information and resources. Advantages of this architecture are:
  - It can be easily configured and installed.
  - All the nodes in the network are capable of sharing resource and information with other nodes present in the network.

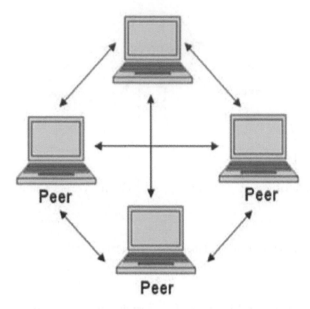

FIGURE 1.5
Peer-to-peer architecture. (Steen & Tanenbaum (2016).)

- Even if any node goes down, it does not affect the whole system.
- Maintaining and building such architecture is comparatively cost effective.

1.2.3 Blockchain as a Distributed System

Blockchain is considered a distributed system because its nodes are connected to one another in order to provide a service, share data, or just store data using the P2P architecture. Blockchain network is constantly checking the peers to see if they all match up, i.e., if everyone (participant of the network) on the system has a copy of the ledger, eliminating the possibility of fraud or illegal tampering. If there is an alteration in one peer, all other peers will be notified, and the altered peer will automatically restore its data to the original state before it was altered (Cornelli et al., 2002).

Therefore, blockchain on a P2P network helps in maintaining the consistency of the distributed ledger because when someone joins the network, a full copy of the blockchain is given to everyone and they check that everything is in order and no fraud has happened on the network. When a new block is created, everyone on the network gets the block and verifies that it hasn't been tampered with and then adds the block to the chain if verified and validated. In case the block has been tampered with, the block gets rejected by the participants of the network and won't be added to the chain (Schlosser et al., 2003). Each user/participant point is called a node. All the nodes in the network create *consensus*. They agree about what blocks are valid and which are not. The identifiers of the nodes are cryptographic keys. You can't get to the actual names of the user/participant (Rauchs et al., 2018).

Note: To successfully make an illegal change in a blockchain, all the blocks on the chain will need to be tampered with, proof of work has to be redone for each block, and the defaulter has to take control of more than 50% of the P2P network to get consensus. All of this is almost impossible to do thereby making the blockchain secured.

1.2.4 Distributed Ledger Technology

Distributed Ledger Technology (DLT) refers to a new and rapidly evolving approach to recording and sharing data (transactions) across multiple data stores (ledgers) that all have the same data records and are managed and controlled by a distributed network of computer servers known as nodes (Natarajan et al., 2017). All participants/nodes can see who is using and modifying the ledger. One way to think about DLT is that it is simply a distributed database with certain specific properties such as real-time sharing of ledgers (Bellare & Yee, 2003).

Blockchain is a particular type of DLT that uses cryptographic and algorithmic methods to create and verify a continuously growing, append-only data structure that takes the form of a chain.

1.2.5 Cryptography

Kryptos, which means "hidden," and Graphein, which means "to write," are the two ancient Greek terms that make up the word cryptography. During a communication process, cryptography entails hiding information from a third party (leakage).

1.2.5.1 Cryptography Terminologies

- **Encryption:** This is the procedure for turning plaintext (ordinary text) to cipher text (random sequence of bits that are not readable).
- **Key:** This is a small piece of data that is required to conclude the cryptographic algorithm's result.
- **Decryption:** This is the inverse process of encryption, and it involves converting cipher text to plaintext.
- **Cipher:** This is the mathematical function, or cryptographic algorithm, that converts plaintext to cipher text.

1.2.5.2 Types of Cryptography

Fundamentally, there are three distinct approaches to perform cryptographic algorithms.

1.2.5.2.1 Symmetric-Key Cryptography

A single key is utilized for both the encryption and decryption processes in encryption approach. The challenge of securely passing the key between the sender and the recipient arises when using a single common key. Secret-Key Cryptography is another name for Symmetric-Key Cryptography.

1.2.5.2.2 Asymmetric-Key Cryptography

This encryption method employs a pair of keys, an encryption key and a decryption key, which are referred to as the public (unique) and private keys, respectively, both of which are generated using the same process. Public-Key Cryptography is another name for asymmetric-key cryptography.

One key is kept private (private key), and the other is shared with the other party (public) in the network. Both keys can be used for both encryption and decryption, i.e., the private key can be used to encrypt, while the public key would be used to decrypt and vice versa. The public key is in the form of a bank account number; it can be shared to other parties while the private key is in form of a password, i.e., only one user (the sender) can know it (Lakshman & Malik et al., 2009).

One of the major parts of asymmetric-key cryptography is digital signatures. Digital signature ensures the people on the network are the real users as well as that the transaction is the real transaction to take place (Elrom et al., 2019). Digital signatures provide integrity to the process; they are easily verifiable and cannot be corrupted. They also hold the quality of non-repudiation, making them similar to the signatures in the real world. The digital signatures ensure that the blockchain is valid and the data is verified and correct.

1.2.5.2.3 Hash Functions

This is a different type of cryptography as it doesn't require keys. It generates a hash value of a defined length from plain text using a cipher. A hash value is like a unique fingerprint of a particular block. Tampering with a particular block means that the fingerprint/hash is going to change and because blocks are linked together, i.e., since block 2 has the previous hash from block 1, it is easy to tell which block has been tampered with. Applying the fingerprint principle to a document is basically what the SHA256 Hash is doing in blockchain by giving each of the blocks a unique identification. The algorithm behind SHA256 was developed by the United States National Security Agency and first published in 2001. It was made to be very secure which is why it is one of the core building blocks of the blockchain. SHA simply stands for Secure Hash Algorithm. SHA256 is a hexadecimal hash, meaning it consists of characters 0–9, A–F. It is 256 bits long, meaning the number of bits it takes up in memory is 256. The 256 bit is equivalent to 32 bytes, or 64 bytes in a hexadecimal string format. SHA 256 algorithm works for any digital document. A very important characteristic of a hash is that it produces the same hash if it's the same data without any tampering.

In blockchain, the block number, the transaction, previous hash and nonce controls the hash of a block. A target is usually set for miners by the blockchain algorithm to accomplish a certain hash, any hash higher than the target doesn't count. Setting a target is just a way to create a challenge for miners to tackle/solve. There is no mathematical or computational reason for setting this target. Only hash below the target set is accepted and then allowed to add block to the network. This concept of finding the right hash by guessing the nonce is called crytographic puzzle.

Note: Nonce is another element of a block. It is usually a large number that miners keep guessing to get hashes below the set target hash and when that happens they get the golden nonce.

Note: Tampering with a block automatically changes its hash and this won't correlate with the hash in other ledgers held by other users and this immediately raises a flag and the transaction won't be verified or validated. This is some security factor in the blockchain. Hash can't be reverse engineered to avoid anyone from getting the complete data.

1.2.5.2.3.1 Characteristics of Hash Function That Benefits Blockchain

1. A hash algorithm is always one way. You can't go from hash to document, you can't restore/reverse engineer the document based on the hash.
2. Hash algorithm has to be deterministic. The same result of the hash with the same document.
3. Hash algorithm has to have fast computation.
4. Avalanche effect means taking the same document and giving out the same hash, except even the tiniest change made to the document changes hash completely.

    Note: Avalanche effect usually happens when the nonce is changed slightly, and this is to prevent people from cheating and having an easy guess to get the target.
5. Collisions should not be possible, i.e., there shouldn't be any case of piracy of documents like making two documents have the same hash.

    Note: Hashing, public-private key pairs, and the digital signatures together constitute the foundation for the blockchain. These cryptographic features make it possible for blocks to get securely linked by other blocks, and also ensure the reliability and immutability of the data stored on the blockchain.

1.2.5.2.4 Smart Contracts

These are simple programs that are kept on the blockchain and can be used to exchange transactions based on the program's criteria. When certain circumstances inside the system are met, a smart contract is a basic computer program that conducts predetermined activities. They can also help with the exchange and transfer of anything of value (e.g., money, content, property, birth certificate). The code for smart contracts is written in high-level languages like Solidity. Compilers such as Solidity or Serpent are used to convert the code to byte code.

1.2.5.2.5 Blockchain Use Cases

Some popular domains where blockchain is used are as follows (Biryukov et al., 2014):

1. Banking
2. Payment and transfer
3. Online music
4. Healthcare
5. Internet of things
6. Law enforcement
7. Voting
8. Real estate

1.2.5.2.6 Blockchain Transaction

In any blockchain:

1. All verified and validated transactions are logged, including the time, date, participants involved, and details of every transaction on the network of everyone on the network.
2. Every node in the network has a complete copy of the blockchain.
3. After solving complex mathematical puzzles, miners verify transactions and maintain and update the ledger.
4. Consensus ensures that the nodes agree on the current state of the ledger and every transaction in it automatically and continuously.
5. If someone tries to corrupt a transaction, the nodes will not reach a consensus and thus will refuse to include the transaction in the blockchain because everyone has the same ledger and transaction state.

1.3 Types of Blockchain/Blockchain Network Types

Table 1.1 compares the two major categories of blockchain network: permissioned and permissionless.

TABLE 1.1

Permissioned Blockchain vs Permissionless Blockchain

S/N	Permissioned	Permissionless
1	Examples are Hyperledger blockchain frameworks	Examples are Bitcoin, Ethereum, etc.
2	All network members are known and given specific roles or access controls such as creating the network, inviting people to the network, accepting people on the network or denying people from accessing the network	There is total anonymity in this type of blockchain
3	Only specific people in the organization can verify and add transaction blocks	Everyone can verify and add a block of transactions to the blockchain
4	Permissioned blockchains are also known as private or consortium blockchains	Permissionless blockchains are also called public blockchains

1. **Public Blockchain**: This type of blockchain network is available to anyone with Internet access. Public blockchains have ledgers that are visible to everyone on the network, but you don't know who is behind any of the transactions; anyone can verify and add a block of transactions to the blockchain, for example, bitcoin and Ethereum. This type of blockchain provides total anonymity. A public blockchain is also known as a permissionless blockchain.

2. **Private**: This type of blockchain network is only available to a small number of people. This network is used to solve specific requirements because it allows only specific people within an organization to verify and add transaction blocks. All network members are identified and assigned specific roles or access controls, such as creating the network, inviting people to join the network, accepting people onto the network, or denying people access to the network. This type of blockchain is mostly used by regulated industries, businesses, and other organizations. Private blockchain can be used to handle identity management issues as you know who is behind every transaction unlike in public blockchain. A private blockchain is also known as a permissioned blockchain because it requires pre-verification of the network's participating parties, who are usually known to each other. Hyperledger blockchain framework is an example here. Private blockchain reduces security risks by ensuring that only the parties who want to transact are included in the transaction, rather than displaying the transaction record to the entire network.

The choice between permissionless (public) and permissioned (private) blockchain depends on the application or use case to be implemented. Most enterprise use cases involve extensive vetting before parties agree to do business with each other. An example is supply chain management which is ideal for a permissioned (private) blockchain. On the contrary, when a network already has trust, facilitating parties can transact without necessarily having to verify each other's identity, like the bitcoin blockchain, a permissionless (public) blockchain is more suitable. Blockchain, in a nutshell, uses mathematical solutions as well as algorithms to create a secure distributed ledger that enables transactions without the need for third parties.

1.4 Existing Blockchain Platforms

1.4.1 Key Terminologies

1.4.1.1 A Protocol

A protocol is a set of rules that guides how participant of a network communicate with each other and agree on things, e.g. TCP, IP, and HTTP. In the crypto world, examples of protocols are bitcoin, Ethereum, and litecoin. These crypto world protocol examples rely on blockchain technology, and they all have their own coin. The coin is an innate asset of the protocol which facilitates the interaction of players, and it is used to reward people for mining and/or purchasing.

1.4.1.2 A Token

A token relies on smart contract built on different protocols.

There are various existing blockchain platforms for different use cases. Some of which are described below.

1.4.1.2.1 Bitcoin

Bitcoin is the first decentralized digital currency and the first blockchain that came into implementation by Satoshi Nakamoto in 2008 and launched in 2009.

Bitcoin is regarded as a *protocol* because it does a lot of dictating and agreement on how the network will function properly/correctly. It dictates how participants would come to a consensus on things. It also dictates how public keys and signatures would be used for authentication. Also, it dictates how agreement on updates to the protocol itself will be done.

Bitcoin doesn't facilitate the concept of creating smart contracts. Therefore, there is no *token* on bitcoin.

Bitcoin uses cryptography to control its management and creation as only about 21 million bitcoins can be created, and possibly by 2140, the last bitcoin will be mined. In other words, the cryptographic algorithm of bitcoin controls the creation of bitcoin and therefore, bitcoin cannot just be created as per our need because of its decentralized nature and its cryptographic algorithm. This is why bitcoin is mined (Nakamoto, 2008). The ledger system in blockchain keeps track of how much bitcoin gets transacted.

Bitcoin is an open-source, P2P cryptocurrency that was developed. The system is based on public-private key technology and the decentralized clearing of payments to allow quasi-anonymous transactions. Bitcoin is an independent currency and it does not belong to any government or legal entity (Luther, 2016). Bitcoin utilizes cryptographic security. However, it can easily be understood by examining a typical transaction. For example, both payment participants have a public and private key. To confirm the ownership of a balance of bitcoin, the payer needs its private key. To identify the payee, the payer should use the payee's public key, which is public/open for everyone/participants in the system. For accepting the transaction as valid, the bitcoin software requests all peers on the network to acknowledge the payment is valid. Once the transaction is verified, all other peers are informed that the balance of the payer was transferred to the payee. To spend the money, the new owner would repeat this process (Franco, 2014).

Bitcoin has several advantages over traditional centralized currency, such as:

1. Bitcoin cannot be confiscated while traditional centralized currency can be confiscated.
2. Bitcoin also avoids capital controls and disproportionate taxation.
3. Anyone who owns bitcoin can have access to the funds, as long as he can connect to the Internet and keeps a copy of the private keys.
4. There are no additional costs to storing bitcoins except the initial setup and the proper securing of a wallet for bitcoin users.
5. Bitcoins are easy to transport. For instance, Bitcoin can be saved in storage media (USB flash drive) or uploaded to the cloud.

   Although Bitcoin has several advantages over other currencies, it is not without its weaknesses (Dhillon *et al.*, 2017; Salau *et al.*, 2021). Since bitcoin is open source, it can easily be replicated and this will give room to make substitutes for Bitcoin. This situation has led to an increase in the number of cryptocurrencies that will compete with each other. Consequently, this might provoke hyperinflation and cause collapse.

1.4.1.2.2 *Ethereum*

Ethereum is another example of a protocol. Ethereum is the most popular protocol for building smart contracts. There's hundreds of token on Ethereum, e.g., TRX and AE. Ethereum is also an online open-source, public, blockchain-based distributed computing platform featuring smart contract functionality to build decentralized applications. Ethereum can also be said to be a decentralized platform that runs on the blockchain with the aid of smart contracts, and this smart contract enables business logic to be implemented. The distributed applications here run exactly as it has been programmed without any possibility of downtime, censorship, fraud, or third-party interference.

Important aspects of Ethereum:

1. The value token of the Ethereum blockchain is called ether.
2. Gas is also an important aspect of Ethereum and is used to pay for the computational services on the Ethereum network.
3. In Ethereum, a smart contract is executed on a decentralized computer called Ethereum virtual machine which lies at the heart of the Ethereum architecture.
4. Ethereum has simple and generalized protocols and easy-to-launch scripting language that makes it a platform thriving with a lot of use cases of decentralized applications.
5. Ethereum as a platform is maintained by Ethereum Developer Community.

1.4.1.2.3 Hyperledger

Brian Behlendorf (Executive Director, Hyperledger) defined Hyperledger as an open-source community of communities to benefit an ecosystem of Hyperledger-based solution providers and users focused on blockchain-related use cases that will work across a variety of industrial sectors. Therefore, Hyperledger can be thought of as software that everyone can use to create one's own personalized blockchain services/solution. Unlike other blockchains, especially the public blockchain which runs on a generalized protocol for everything that runs on its network, Hyperledger can be personalized to suit one's need.

Hyperledger began as a small project in 2015 by various developers in different sectors like finance, supply, and data management to make blockchain technology accessible to the world. With this goal in mind, developers started testing interactions between applications and secure blockchain networks. They then realized that public blockchain has a restriction of requiring each peer to execute every transaction and run "consensus" at the same time, thereby making public blockchain not scale-able. Public blockchain also does not support private and confidential transactions.

Note: Private and confidential contracts means that only the major parties/parties directly affiliated with the deal/transaction are updated on the ledger and notified. Third parties do not need to know the details of every transaction, unlike the public blockchain where all/every ledger will be updated about every transaction on the network.

Hyperledger is an open-source collaborative effort/project created to advance cross-industry blockchain technologies. It is under the Linux Foundation. JP Morgan and IBM are increasing the awareness of Hyperledger by contributing.

Hyperledger is different from other existing blockchain platforms because of the peers which create three distinct roles on the network that runs on separate run time. The roles are as follows:

1. **Endorser:** Endorsers are peers responsible for simulating transactions put in the network and preventing unstable and non-deterministic transactions.

   Note: Committers are peers that may or may not be endorsers depending on the network restrictions stated.

2. **Committer:** Committers are peers responsible for appending validated transactions to their specific ledger once returned by the consenters.

3. **Consenter:** Consenters are peers responsible for:
   - Network's consensus service
   - A collection of consensus service nodes will order transactions into blocks according to the network's chosen ordering implementation
   - Running the consensus algorithm on the network

   Note: Consenters can be referred to as the bodyguard of the network who validates every transaction deciding whether or not it should be added to the ledger.

Hyperledger is a group of open-source projects focused on distributed ledger technologies, hosted and launched by the Linux Foundation in December 2016. Hyperledger consists of ten major projects. Five of them are distributed ledger frameworks as follows:

a. Hyperledger Fabric, which is the most used blockchain framework after Ethereum

b. Hyperledger Sawtooth, which is designed for IoT-based use cases

c. Hyperledger Iroha, which allows institutions to share data and manage customer identity using the Know Your Customer (KYC) service

d. Hyperledger Burrows, which accepts Ethereum-based smart contract code

e. Hyperledger Indy, which supports the user-controlled exchange of verifiable claims about identifying information.

The other five projects are tools that support these frameworks. The Hyperledger tools are auxiliary software used for deploying and maintaining Blockchain applications, examining the data on the ledgers as well as tools to design and extend Blockchain networks (Glaser, 2017).

The Hyperledger tools available are Hyperledger Cello, Hyperledger Explorer, Hyperledger Composer, and Hyperledger Caliper.

Hyperledger Cello provides a toolkit that deploys Blockchain as a Service. In other words, Hyperledger Cello allows Blockchain deployment on the cloud. Operators can create and manage Blockchain through a dashboard, and users can obtain a blockchain instance immediately.

Hyperledger Explorer is a tool for visualizing Blockchain operations. It allows anyone to explore the distributed ledger without compromising his or her privacy. Hyperledger explorer includes a web browser called Playground. The Playground is an environment that allows building and testing Blockchain business networks. Playground runs in a Docker container and can be installed either with Hyperledger fabric validating peer network or with browser-only.

Hyperledger Caliper is a blockchain benchmark tool that allows users to measure the performance of a specific blockchain implementation with a set of predefined use cases. Hyperledger Caliper produces reports containing several performance indicators (e.g., transactions per second, transaction latency, and resource utilization). These reports can be used in deciding if blockchain implementation is suitable for a user's specific need.

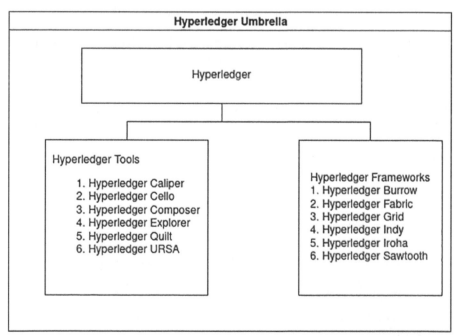

FIGURE 1.6
Hyperledger frameworks and tools.

The purpose of Hyperledger is to provide an alternative solution to the cryptocurrency-based blockchain model, and focus on developing blockchain framework and modules for global enterprise solutions support. Figure 1.6 gives a pictorial view of the Hyperledger umbrella showing its frameworks and tools as explained above.

1.5 Consensus Mechanism

This is a mechanism that governs how nodes on the distributed network agrees on how to keep adding blocks and which block to add across the whole network. In other words, Consensus in the network refers to the process of achieving agreement among the network participants as to the correct state of data on the system (Vukolić, 2015). A consensus algorithm for blockchain does two things which are regarded as key objectives of consensus algorithm:

- It assures that all nodes in the network, i.e., rival chains, have the same data on the ledger, including lags between nodes that are far apart.
- It protects the network from attacks by preventing hostile actors from changing the data, which is mostly how new blocks are created.
  To achieve these key objectives, the process of consensus follows four steps:
  1. Each node creates the transactions it wants to record.
  2. The data is shared between the nodes.
  3. Consensus is established on the order of valid transactions.
  4. Nodes update the transactions to reflect the consensus result.

The goal is to get to process the four steps as quickly as possible without breaking consensus.

There are various consensus algorithms such as proof of work, PoS, and Byzantine fault-tolerant replication (Gramoli, 2020). Bold new models are also proposed, such as PoS and Proof of Elapsed Time (PoET) and variations of Practical Byzantine Fault Tolerance (PBFT) appear as viable alternatives.

The applicability and efficacy of a consensus process can be judged by three important features as stated by Baliga (2017).

a. **Safety:** A consensus protocol is determined to be safe if all nodes produce the same output and the outputs produced by the nodes are valid according to the rules of the protocol. This is also referred to as the consistency of the shared state.

b. **Liveness**: A consensus protocol guarantees liveness if all non-faulty nodes participating in consensus eventually produce a value.

c. **Fault Tolerance**: A consensus protocol provides fault tolerance if it can recover from the failure of a node participating in consensus.

Prominent blockchain platforms use a variety of consensus algorithms such as:

a. **Proof of Work**: Proof of work is a consensus method that makes the generation of new blocks take longer. In bitcoin, calculating the requisite proof of work and adding a new block to the chain takes roughly 10 minutes. This makes it difficult to tamper with the block because if one block is tampered with, the proof of work for all subsequent blocks must be recalculated, which is nearly impossible, hence making the blockchain system more secure. Proof of work is a method of distributing some kind of asset in a blockchain system randomly (Bocek et al., 2017).

b. **PoET**: Intel SawtoothLake, IntelLedger, or Intel SawtoothLake is a blockchain platform developed by Intel and subsequently open-sourced for use by the community. The project is officially now under Linux Foundation's HyperLedger project as a proposal for further development. IntelLedger uses a consensus algorithm, designed by Intel, called Proof of Elapsed Time (PoET) intended to run in a Trusted Execution Environment (TEE), such as Intel's Software Guard Extensions (SGX). PoET uses a random leader election model or a lottery-based election model based on SGX, where the protocol randomly selects the next leader to finalize the block. The random leader election algorithm uses this model to deal with untrusted nodes and open-ended participation of nodes in the consensus algorithm. For the consensus to work correctly, it has to randomly distribute the leader election among all available participating nodes and it needs a secure way for other nodes to verify that a given leader was correctly selected without any scope for manipulation. This is achieved using the TEE to guarantee the safety and randomness of electing a leader. Leader election works as follows. All validating or mining nodes have to run the TEE using Intel SGX. Each validator requests a wait time from the code running inside the TEE. The validator with the shortest wait time wins the lottery and can become the leader. The functions within the TEE are designed such that their execution cannot be tampered with by external software. When a validating node claims to be a leader and mines a block, it can also produce proof generated within the TEE that other nodes can easily verify. It has to prove that it had the shortest wait time and it waited for a protocol designated amount of time before it is allowed to start mining the next block. The randomness in generating wait times ensures that the leader role is randomly distributed among all validating nodes. The only drawback of this algorithm is the reliance on specialized hardware.

c. **Byzantine Fault Tolerance**: Hyperledger Fabric, which is the most popular Permissioned blockchain platform being developed by the Linux Foundation, provides a flexible architecture with a pluggable consensus model. Fabric is designed for consortium where the group of participants in the consortium is not only known but their identities are registered and verified with a central registry service running within the system. It also supports smart contracts on the blockchain, also known as Chain code. Hyperledger currently supports the popular PBFT algorithm and its variation SIEVE that can handle non-deterministic chain code execution. Current proposals are considering Crash Fault Tolerance (Schwartz et al., 2014), which is a variation of Paxos with Byzantine Fault Tolerance built-in, as an alternative consensus algorithm for future versions. The PBFT algorithm proposed by Miguel Castro and Barbara Liskov was the first practical solution to achieving consensus in the face of Byzantine failures. It uses the concept of a replicated state machine and voting by replicas for state changes. It also provides several important optimizations, such as signing and encryption of messages exchanged between replicas and clients, reducing the size and number of messages exchanged, for the system to be practical in the face of Byzantine faults. This algorithm requires "3f+1" replicas to be able to tolerate "f" failing nodes. This approach imposes a low overhead on the performance of the replicated service. The authors report a 3% overhead for a replicated network file system service that they conducted their experiments on. PBFT, however, has only been scaled and studied to 20 replicas (Mazieres, 2015). Its messaging overhead increases significantly as the number of replicas increases. SIEVE consensus protocol is designed to handle non-determinism in chain code execution. When non-determinism is present within the chain code, it can produce different outputs when executed by different replicas in a distributed network. SIEVE handles transactions that are usually deterministic, but which may occasionally yield different outputs. The SIEVE protocol treats the chain code itself like a black box. It initially executes all operations speculatively and then compares the outputs across replicas. If the protocol detects a minor divergence among a small number of replicas, the diverging values are sieved out. If the divergence occurs across several processes, then the offending operation itself is sieved out (Lamport et al., 2019).

Other consensus algorithms are Cross-Fault Tolerance, Federated Byzantine Agreement, Ripple Consensus Protocol Algorithm (Schwartz et al., 2014), Stellar Consensus Protocol (Mazieres, 2015), and PoS (Liu et al., 2016).

References

Attaran, M., & Gunasekaran, A. (2019). Blockchain and cybersecurity. In: *Applications of Blockchain Technology in Business* (pp. 67–69). Springer, Cham.

Baliga, A. (2017). Understanding blockchain consensus models. *Persistent*, 4, 1–14.

Bellare, M., & Yee, B. (2003). Forward-security in private-key cryptography. *In Cryptographers' Track at the RSA Conference* (pp. 1–18). Springer, Berlin, Heidelberg.

Bocek, T., Rodrigues, B. B., Strasser, T., & Stiller, B. (2017). Blockchains everywhere- a use- case of blockchains in the pharma supply-chain. *In 2017 IFIP/IEEE Symposium on Integrated Network and Service Management (IM)* (pp. 772–777). IEEE.

Biryukov, A., Khovratovich, D., & Pustogarov, I. (2014). Deanonymisation of clients in Bitcoin P2P network. *In Proceedings of the 2014 ACM SIGSAC Conference on Computer and Communications Security* (pp. 15–29). Scottsdale, Arizona, USA.

Cornelli, F., Damiani, E., Vimercati, S.D., Paraboschi, S., & Samarati, P. (2002). Choosing reputable servents in a P2P network. *In Proceedings of the 11th International Conference on World Wide Web* (pp. 376–386). Scottsdale, Arizona, USA.

Dhillon, V., Metcalf, D., & Hooper, M. (2017). The Hyperledger project. In: *Blockchain Enabled Applications* (pp. 139–149). Apress, Berkeley, CA.

Elrom, E. (2019). Build Dapps with Angular: Part I. In *The Blockchain Developer* (pp. 349–394). Apress, Berkeley, CA.

Franco, P. (2014). *Understanding Bitcoin: Cryptography, Engineering and Economics.* John Wiley & Sons, Hoboken, NJ.

Glaser, F. (2017). Pervasive decentralisation of digital infrastructures: A framework for blockchain enabled system and use case analysis. HICSS.

Gramoli, V. (2020). From blockchain consensus back to Byzantine consensus. *Future Generation Computer Systems*, 107, 760–769.

Lakshman, A., & Malik, P. (2009). Cassandra: Structured storage system on a p2p network. *In Proceedings of the 28th ACM Symposium on Principles of Distributed Computing* (pp. 5–5).

Lamport, L., Shostak, R., & Pease, M. (2019). The Byzantine generals problem. In: Malkhi, D. (ed.) *Concurrency: The Works of Leslie Lamport* (pp. 203–226).

Liu, S., Viotti, P., Cachin, C., Quéma, V., & Vukolić, M. (2016). XFT: Practical fault toler-ance beyond crashes. *In 12th USENIX Symposium on Operating Systems Design and Implementation* (OSDI 16) (pp. 485–500). Scottsdale, Arizona, USA.

Lopes, J., & Pereira, J. L. (2019). Blockchain projects ecosystem: A review of current technical and legal challenges. *In World Conference on Information Systems and Technologies* (pp. 83–92). Springer, Cham.

Luther, W. J. (2016). Bitcoin and the future of digital payments. *The Independent Review*, 20(3), 397–404.

Mazieres, D. (2015). The stellar consensus protocol: A federated model for internet-level consensus. *Stellar Development Foundation*, 32, 1-45.

Nakamoto, S. (2008). Bitcoin: A peer-to-peer electronic cash system. *Decentralized Business Review*, 21260.

Natarajan, H., Krause, S., & Gradstein, H. (2017). *Distributed Ledger Technology and Blockchain.* World Bank, Washington, DC.

Rauchs, M., Glidden, A. L., Gordon, B., Pieters, G., Recanatini, M., Rostand, F., Vagneur, K., & Zhang, B. Z. (2018). Distributed ledger technology systems: A conceptual framework. Available at SSRN 3230013.

Salau, A. O., Marriwala N., & Athaee M. (2021). Data security in wireless sensor networks: Attacks and countermeasures. *Lecture Notes in Networks and Systems* (vol. 140, pp. 173–186). Springer, Singapore. doi: 10.1007/978-981-15-7130-5_13

Schlosser, M., Condie, T., & Kamvar, S. (2003). Simulating a file-sharing p2p network. *In 1st Workshop on Semantics in Grid and P2P Networks*. Stanford InfoLab. Scottsdale, Arizona, USA.

Schwartz, D., Youngs, N., & Britto, A. (2014). The ripple protocol consensus algorithm. *Ripple Labs Inc White Paper*, 5(8), 151.

Steen, M. V., & Tanenbaum, A. S. (2016). A brief introduction to distributed systems. *Computing*, 98(10), 967–1009.

Vukolić, M. (2015). The quest for scalable blockchain fabric: Proof-of-work vs. BFT replication. *In International Workshop on Open Problems in Network Security* (pp. 112–125). Springer, Cham.

Yaga, D., Mell, P., Roby, N., & Scarfone, K. (2018). Blockchain technology overview. arXiv: Cryptography and Security.

2

Current Research Trends and Application of Blockchain in Healthcare and Medical Systems

Avita Katal

School of Computer Science, University of Petroleum & Energy Studies (UPES)

CONTENTS

DOI: 10.1201/9781003193425-2

2.1 Introduction

Blockchain is the core mechanism that powers a wide range of digital currencies. Blockchain is a series of blocks that holds information with electronic signatures in a decentralised and distributed network. Decentralisation, flexibility, traceability, and accountability are some of the blockchain features that make payments more safe and tamper-proof. Blockchain technology may be utilised in areas other than cryptocurrencies, such as services related to finance and social, management of risk, and healthcare facilities. Research has been conducted to examine the opportunities that blockchain presents in various application fields. In contrast to previous techniques, blockchain allows for peer-to-peer transmission of digital assets without the need of middlemen. Blockchain was first developed to support the well-known cryptocurrency Bitcoin. Nakamoto proposed Bitcoin in 2008 and implemented it in 2009 (Nakamoto, n.d.). Blockchain is essentially a series of blocks that uses a public ledger to store all committed transactions (Salah et al., 2019). When additional blocks are added to the chain, it continues to expand indefinitely. Blockchain operates in a decentralised environment supported by many key technologies such as digital signatures, cryptographic hashes, and distributed consensus methods. Although Bitcoin is the most well-known use of blockchain, it may be used for a wide range of purposes other than currency. Since it enables payments can be made without participation of a bank or an agent, it is utilised in a number services, such as electronic assets, remittance, and electronic transaction. It has developed on its own personality and has invaded a range of businesses, such as banking, medical, administration, industry, and transportation. The blockchain has the ability to break new ground and revamp a number of uses, such electronic mainstream press transfer, distant service provision, systems such as desktop moving to findings of information, and dispersed certification. Other applications of blockchain include distributed resources (energy generation and distribution), fundraising, digital polling, access control, and controlling public publications. Regardless of the fact that blockchain technology has the alternative to transform many of contemporary online media, it is still in its early stages, and it does have certain technological limitations. For blockchain-based platforms, scalability is a major problem.

2.2 Background Concepts on Blockchain Technology

2.2.1 Consensus Algorithms

The origins of consensus algorithms may be traced back to the emergence of distributed systems, where they were governed by transitions within these networks. A consensus algorithm is a technique which enables all Blockchain network participants to agreement on the existing iteration of the shared database. Consensus methods, in this way, promote reliability in the Blockchain network and develop trust between unidentified neighbours in a decentralised computational system. Blockchain technology employs a number of consensus algorithms. They are classified into two types: proof-based consensus algorithms and voting-based consensus algorithms (Katal et al., 2021).

2.2.2.1 Proof-Based Consensus Algorithms

An evidence-based consensus algorithm's core theory is that among numerous nodes entering the network, the node with the greatest proof prevails and is granted the right to add a new block to the chain. There have been several types of proof-based consensus algorithms, such as those based on Proof of Work, Proof of Stake, or a mix of the two, as well as a variety of additional types that aren't the two major ones.

2.2.2.2 Voting-Based Consensus Algorithms

To construct the voting-based consensus process, the nodes in the verifying network must be known and configurable. This makes communication simpler for them. This is the most significant distinction between proof-based consensus methods and those that allow nodes to join and depart the verifying network on a regular basis. Furthermore, all nodes in the network must jointly validate transactions or blocks in addition to maintaining the ledger via a voting-based consensus method. They can consult with others before determining whether or not to incorporate the suggested blocks in their chain. The use of voting-based consensus methods is analogous to the use of classical fault tolerance approaches for distributed systems (Andreev et al., 2018). As a result, voting-based consensus should be built to withstand circumstances in which one or more nodes fail, crash, or subvert.

2.2.2 Types of Blockchain

Private and public blockchains are the two main types of blockchains. There are, however, several variants, such as Consortium and Hybrid blockchains. Figure 2.1 depicts how blockchain works.

Public Blockchain: It is a public ledger with no permissions. Anyone with access to the Internet would be able to join the blockchain network. A user (also known as a node) who is a member of the public blockchain can access data, verify transactions, and mine for the next block. One of the most often used implementations of decentralised blockchains is cryptocurrency exchange. The majority of the time, shared blockchain is safe as long as users follow security protocols (Salau et al., 2021).

Private Blockchain: It is a closed-network permissioned blockchain. They are typically used for a small group or business. Security, authorisations, and compatibility are only a couple of the critical mechanisms under the control of a governing entity. A private blockchain can be used to execute any specific operation (such as voting, supply chain control, and asset ownership).

Consortium Blockchain: It is a partially decentralised blockchain network that is run by a number of different companies. More than one organisation may act as the regulator for mining and information sharing for this form of blockchain.

Hybrid Blockchain: It is a blockchain network that brings together private and public blockchain networks. The functionality of all blockchains is included in this situation. On the hybrid network, users can keep track of who has access to which blockchain data. Just a handful of the blockchain's chosen records can be made public, while the others are kept private in the private network. However, for authentication purposes, users can post this on the public blockchain. The hashing power of public blockchains grows, and extra peers participate in the validation. This increases the security and transparency of the blockchain network.

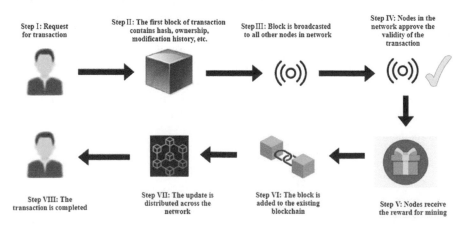

Step I: Request for transaction

Step II: The first block of transaction contains hash, ownership, modification history, etc.

Step III: Block is broadcasted to all other nodes in network

Step IV: Nodes in the network approve the validity of the transaction

Step VIII: The transaction is completed

Step VII: The update is distributed across the network

Step VI: The block is added to the existing blockchain

Step V: Nodes receive the reward for mining

FIGURE 2.1
Working of blockchain.

2.2.3 Smart Contracts

Nick Szabo advocated encoding a variety of contractual obligations in software and hardware to make contract breaches costly (Wang et al., 2019). The major advantage of using smart contracts over commercial procurement is the control layer's quickness. Adopting smart contracts, for example, makes it much easier to keep track of previous contractual partners' behaviour. A smart contract is an intelligent agent. In other words, it is computer software that can make decisions if certain conditions are met. The complexity of the transaction that an agent is meant to accomplish determines its intelligence. Contracts can be relatively basic transactions that are done in seconds or minutes, or they can be highly intricate and lengthy transactions that include discussions and tens of pages of written language with specific rights and obligations that may take hours or months to execute. Smart contracts are presently categorised as simple transactions. The author in CoinDesk (n.d.) divided smart contracts into two categories: smart contract code and smart legal contract. Smart contract code is defined as "code that is recorded, validated, and executed on a blockchain" (CoinDesk, n.d.). The functioning of a smart contract is entirely reliant on the programming language used to express the contract and the blockchain's capabilities. The term "smart legal contract" refers to software that can augment or replace legal contracts. Rather than technology, the functioning of this smart contract is determined by legal, political, and business organisations. A smart contract has a balance, safe storage, and computer code. The contract's state is made up of the contract's storage and balance. When the contract is performed, the information on the blockchain is stored and updated. Each contract will be given a distinct 20-byte address. Once the contract code has been put on the blockchain, it cannot be changed. A contract can be performed by simply sending a transaction to the address specified in the contract. This operation will then be carried out by each agreement node (also known as miners) on the system in order to reach an agreement on its finish. The status of the contract will be updated as needed. Smart contracts are built up of a set of procedural rules and logic known as Scenario-Response logic. Figure 2.2 illustrates the functioning mechanism of a smart contract. Until the parties consent on a smart contract, it is typically linked to a blockchain in the format of machine code and is stored in the chain after already being transmitted over the peer-to-peer (P2P) system and validated by the nodes.

2.2.4 Characteristics of Blockchain

- **Decentralisation**: Every action in a traditional centralised transactions handling platform should be certified by a trusted centralised entity. A blockchain transaction may be carried out between any two identities (P2P) without the need for central authority verification. As a consequence, blockchain has the ability to drastically cut hosting

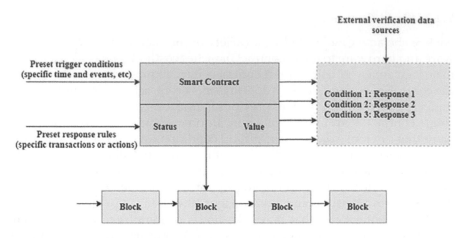

FIGURE 2.2
Operating mechanism of smart contracts.

costs (both planning and production expenses) while also removing compatibility concerns at the centralised computer.

- **Persistence**: Because each network transaction must be verified and stored in blocks spread throughout the whole system. Every sent transaction will be verified and reviewed by a large number of other peers. As a result, any forgery may be detected immediately.

- **Anonymity**: A created address may be used by any user to connect to the blockchain network. Furthermore, in order to conceal their identity, a person may create a large number of addresses. There is no longer a centralised body in charge of protecting users' private information. This approach assures that the blockchain's transactions remain anonymous. It should be highlighted that blockchain cannot guarantee anonymity.

- **Auditability**: Users may quickly examine and trace past records by connecting to a node in the decentralised system as every blockchain operation is verified. Each operation in the Blockchain network may be repeatedly traced back to previous transactions. It increases the transparency and traceability of data stored on blockchains.

2.3 Applications of Blockchain in Healthcare

Originally intended for its most well-known applications in economics and cryptocurrency, blockchain technology (BCT) is currently being used in a variety of other sectors, including biomedicine (Kuo et al., 2017). The capability of blockchain technology could even be noticed in the areas of medication, genome sequencing, telehealth, tele-monitoring, e-health (Adamu et al., 2021), neurobiology, and individualised public healthcare software, due to its method for sustaining and storing the data set in whom the customers can connect through numerous types of accounts.

2.3.1 Electronic Health Records

Healthcare professionals, clinics, and medical apparatus have pushed a growth in the automation of professional medical information over the last decade, because digitalisation of this information provides for quicker access and exchange. However, EHRs are never meant to manage lifelong data throughout many organisations, and individuals leave their data spread throughout numerous organisations when life experiences divide people through one provider's data into the other; as a result, they lose simple access to earlier data (Mandl et al., 2001; Brandon et al., 1991). Many academics have mentioned blockchain technology in the context of maintaining EHRs in reaction to an urgent need for a new strategy to dealing with EHRs as a manner which promotes consumers to engage with their present and past health records.

Primary data is generated when a patient interacts with their physicians and specialists. This information includes a health records, a present issue, and additional physiologic information. With the basic data obtained during first phase, an EHR is built for each individual. Additional types of medical information, including those produced by nursing home care, diagnostic imaging, and medication information, are incorporated in EHR. Single client in possession of critical EHR, with access restriction limited to the property's owner exclusively. Organisations wishing to gain access to certain confidential material must submit a request, which should subsequently be routed to the EHR owner, who chooses who will be allowed access. Databases and cloud storage spread information, while blockchain allows for extreme confidentiality to ensure targeted true user access. End users that seek to obtain access to a safe and secure care delivery system that has been authorised by the owner include ad hoc clinics, community care centres, and hospitals. Figure 2.3 shows the electronic health record mechanism using blockchain.

A model called "MedRec" (Azaria et al., 2016) is a blockchain-based distributed records management tool for EMRs. Because of advances in technology, individuals may now view personal health data in real time among many healthcare providers and processing transforms. Blockchain technology, in particular, is used to facilitate identity, confidentiality, responsibility, and data exchange.

Medical data sharing frequently confronts major constraints during EHR adoption, such as information theft, information origin, monitoring, and secured data monitoring on medical information. Keeping these restrictions

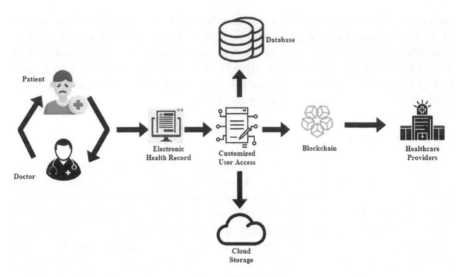

FIGURE 2.3
Electronic health record mechanism using blockchain.

in mind, Xia et al. (2017) created MeDShare, a stable and sound blockchain system for exchanging medical information between network participants. It could perhaps be utilised for transmitting medical information and preserve details of patients across cloud providers, institutions, and healthcare insurance research groups, with tighter data integrity, individualised quality guarantee, and less possible dangers to data privacy and security.

EHRs often include extremely sensitive and essential patient data that is regularly exchanged among physicians in order to offer effective diagnosis and treatment. The physician's treatment may be threatened throughout the preservation, transmission, and distribution of this very delicate patient data around many organisations, which can pose serious dangers to healthcare and the ability to keep an up-to-date health record. Patients suffering from chronic diseases are more likely to be exposed to such hazards as a result of a lengthy record of pre-and-post therapy, follow-ups, and recovery methods. As a result, keeping the latest details of the patient has become critical to ensuring effective therapy. To overcome such constraints, Dubovitskaya et al. (2017) designed a blockchain-based architecture for organising, preserving, and exchanging cancer patients' electronic medical information. For accessing, managing, and storing protected patient information, they used permissioned blockchain technology.

2.3.2 Clinical Research

In clinical trials, a variety of difficulties such as privacy protection, data security, information sharing, data collection, patient registration, and other issues may arise. Blockchain, as the next generation of the internet, has the potential to deliver practical answers to these challenges. It is being used by healthcare researchers to address these challenges (Nugent et al., 2016). Blockchain technologies, supported by machine learning and artificial intelligence (AI), will soon take over the healthcare business. Permissioned Ethereum, a technology that offers smart contract capabilities in blockchain (Wood, 2014), is utilised in tandem to clinic-based data management systems in the study presented by Nugent et al. (2016). The major objective of the research was to tackle the problem of patient registration. The authors stated that Ethereum led in faster transaction over bitcoins, and therefore the outcome validated the use of Ethereum smart contracts for system of management of data transparency in clinical trials. As a result, client recruitment via blockchains is one of the recognised areas of application in clinical research. Mehdi Benchoufi et al. (2017) performed another study in which a framework was created to get informed permission from patients for tracking and keeping data in a safe, publicly verifiable, and unfalsifiable manner. They created its process using blockchain technology.

2.3.3 Medical Fraud Detection

Pharmaceutical medication supply chain management is a significant use of blockchains in the medical business. Supply management is a critical problem to protect in many industries, but it is more important in healthcare owing to its rising complexity. This is due to the fact that every disruption in the healthcare supply chain has an impact on a patient's well-being (Clauson et al., 2018). Because there are so many moving components and individuals involved in supply chains, they are susceptible and have gaps for fraudulent attacks. Blockchains, through improving information visibility and enhancing service tracking, also provide a stable and sound framework to address this problem and, in some cases, stop theft from happening. The blockchain is difficult to manipulate because a transaction on the blockchain can be validated and modified via a smart contract.

2.3.4 Pharmacy Research

The drug company is among the most rapidly growing, and it is a pioneer in healthcare provision. The pharma business not only helps to bring novel and genuinely valuable treatments to the marketplace, but it also ensures the safety and legitimacy of prescription drugs and medicines sold to end consumers. Furthermore, the pharmaceutical industry assists in the assessment (Mettler, 2016) and processing of safe medicines, resulting in a faster patient recovery. In most situations, pharma firms encounter difficulties in tracking their goods on time, which can pose serious dangers by allowing counterfeiters to disrupt manufacturing or infiltrate phoney medications into the system. As a result, the manufacture and sale of counterfeit pharmaceuticals has emerged as a serious and global health issue, particularly in developing nations. Blockchain technology may be a good fit for analysing, tracking, and ensuring the production process of potential medications during design and distribution. Hyperledger recently established a counterfeit medicine initiative that would use blockchain technology as a primary tool for checking and combating the manufacture of counterfeit pharmaceuticals. In the case of delivering trustworthy and legitimate medications to patients, there is an essential necessity to watch, analyse, and ensure the entire system of generating and distributing pharma medications via the application of digital technology worldwide, mainly in developing nations, for the health of customers. A digital drug control system (Plotnikov & Kuznetsova, 2018) might be a long-term answer for the prevention of counterfeit medicines in this area. Big pharmaceutical companies established a collaborative pilot project for new medication inspection and assessment using a blockchain-based digital drug control system.

2.3.5 Telemedicine

Telemedicine, which provides a trusted level among healthcare professionals and consumers, is another sector of healthcare that could profit from blockchain technology. A blockchain-based healthcare system can verify expert identity and message security, offer openness and provenance, and motivate participants to perform well by providing reward metrics such as character rankings. The successful development of such platforms might lead to the establishment of a worldwide health exchange in the long term, effectively regulating global supply and consumption for medical research and finding. Additionally, such a network can contain software solutions such as Artificial Intelligence (AI) if the AI models are appropriately validated and managed, and their effectiveness is continuously monitored using some proxy metrics. Remote diagnostic services are expected to have been at the cutting edge of blockchain use in telehealth. Detection solutions relying only on qualitative analysis of health information in the absence of a client are expected to be the one to successfully use blockchain technology. This expectation is supported by a high number of startups in this field.

2.3.6 Neuroscience

The number of blockchain-related news and research is quickly rising, and neurology is certainly involved. Modern brain technology seeks to develop a novel model that eliminates physical connection with surroundings and enables individuals to connect with gadgets and information via cognitive instead of mechanical commands. These neural systems are capable of reading brain activity patterns that are converted into orders for controlling outside equipment, and information from either a person's brain activity is used to predict a person's personal current mental condition. Complex algorithms and huge data can be used with blockchain theory to record such brain signals on the neural interface. Neurogress (The Hindu BusinessLine, n.d.) is one of the organisations that has verified its use of blockchain technology. The gadget protects and keeps personal information private. As a result, any aberrant behaviour will be easily traceable. It is obvious that blockchains are a sort of IT with a wide range of possible applications, including brain augmentation, brain modelling, and brain thinking. A complete human brain must be digitised, which demands the usage of a medium to archive it, and here is where blockchain technology enters the picture once more. One idea is to store mindfiles, which may be used as storage building blocks in personal thought chains and shared in a P2P network file system. This perspective on blockchain is characterised as an input-processing-output computer architecture with a number of qualities that enable AI technology, individual development, and its inclusion. Blockchain requires an interconnected network with machines joining at timestamp intervals to validate the origins and authenticity of a database.

2.4 Blockchain-Based Supply Chain Management

In the medical sector, the importance of the supply chain cannot be emphasised. To guarantee optimal and intended utilisation, effective monitoring and tracking are necessary from raw materials and manufacturing to various phases of storage and distribution. Counterfeit medicines have been a major problem in recent years. End consumers and other participants in the supply chain should be able to authenticate the constituents of a medicine. There are various weak points in the supply chain in which the drug can be tampered with or a fake drug can be administered since there is no effective surveillance system in play. To address this rising problem, new laws have been proposed that require all stakeholders in the pharmaceutical supply chain to establish a comprehensive method for tracking and tracing pharmaceutical supplies that travel through them. Since this content should be held in an accessible but secured system usable to many entities, blockchain is an excellent solution for this necessity for monitoring and surveillance. Blockchain technology transactions are an especially important monitoring tool in healthcare supply chain management for accessing into the entire process of medication and medical product transportation (Dujak & Sajter, 2019). Even though all activities are stored on the chain as well as each node on the blockchain keeps a record of the transaction, it is simple to validate the source of the medication, the merchant, and the delivery. Moreover, blockchain enables healthcare professionals and clinicians to verify and confirm the qualifications of distributors (Narayanaswami et al., 2019). Shops and care providers make sure that legal drugs, try to speak to the people who require them the most with greater insight into the supply chain provided by a correct and timely authentication procedure. In this view, it ensures for creating a trusted web of providers, allowing health administrators to safeguard clients from dishonest vendors. Moreover, blockchain technology has the ability to enhance predictive analytics, fraud detection, and transactions considerably. As a result, many systems based on blockchain have been developed to trace the pharmaceutical supply chain. The MediLedger project (MediLedger, n.d.) is attempting to create a network that is open to the pharmaceutical industry. The system is a permissioned blockchain for pharma supply chain participants. The idea adheres to the track and trace requirements. Ambrosus (Ambrosus.Io, n.d.) is one of the most ambitious initiatives in this sector, with its flagship product. Companies may develop their own supply chain systems on top of AMB-net. MODsense T1 by Modum (n.d.) is a temperature sensor that monitors conditions in the pharmaceutical supply chain, assisting in meeting regulatory criteria for Good Distribution Practice of medical goods for human consumption. A blockchain is used to store sensor data and other digital documents. Several additional blockchain-based supply chain technologies mention the pharmaceutical sector as one of their major application areas.

2.4.1 Related Work

The pharmaceutical supply chain is notable among the health supply chain's affected regions. Every industry's largest difficulty is effective supply chain management. However, there is an added danger and complexity in healthcare since a faulty supply chain might jeopardise the patient's safety. Increased technological use and globalisation in an industry with various stakeholders have resulted in a convoluted health supply chain. Some of the models proposed by researchers are mentioned below.

Jamil et al. (2019) capture actions on the network in attempt to create a sustainable healthcare framework with a medicine supply chain. A smart contract is being developed to provide time-limited access to electronic medication data as well as patient electronic health information. They utilised Hyperledger Caliper as a measuring tool to assess the effectiveness of the intended platform in respect of transactions per second, transaction delay, and resource utilisation.

Liu et al. (2021) proposed a five-layer Blockchain and Internet of Things-based intelligent tracking and monitoring technology to give a decentralised transparency solution in the drug supply chain. A possible approach for the pharmaceutical sector to accomplish blockchain concept, implementation, deployment, and evaluation is described using the five-layer blockchain system design. Furthermore, three major enablers are introduced: Internet of Things (IoT)-based medicine identification administration, on-chain and off-chain processes, and smart contract-enabled pharmacological services.

Huang et al. (2018) presented Drugledger, a scenario-oriented blockchain solution for drug traceability and control that recreates the whole service design by dividing service providers into three separate service parts and protects the validity and confidentiality of tracking data. Furthermore, Drugledger was able to properly prune its store, resulting in a blockchain storage that was ultimately stable and acceptable. It also enhanced Unspent Transaction Output (UTXO) process is used to create algorithms that represent real-world drug supply chain logic.

The above-mentioned papers demonstrate the use of blockchain in drug supply chain. They are either incomplete or cannot be used in the present scenario as many of these models don't provide the provision of tracing the real-time location. These models did not trace the location of the product in shipment stage. The proposed approach tracks the real-time location of the product and stores it in the blockchain for verification.

2.4.2 Proposed Framework

To implement blockchain technology into the pharmaceutical supply chain system, one first must comprehend how well the blockchain ledger operates. An encryption secure key pair serves as a constructed identity mechanism in blockchain. Those credentials are being used to assign a unique action for each network member. The original identities of the individuals are concealed, and they are only known through these keys. A key pair includes no information about the participant, but it can be linked with extra information. It is also important to pick a certain blockchain network for recording events. A permissioned blockchain is the preferred option inside the setting of the pharmaceutical industry. The next step is to utilise a certain blockchain network to preserve the transaction record, although this is entirely up to the developer's discretion. Figure 2.4 shows the proposed model for the drug supply chain based on blockchain.

Only trustworthy entities are allowed to enter the system under the suggested paradigm. On the backend, there is indeed a permissioned blockchain that keeps all necessary events and cannot be changed once data is entered into it. Aside from that, a user-friendly application is made that participants will use to perform blockchain transactions. In the supply chain, there are five entities that complete the entire supply chain. These are raw material provider, manufacturer, distributor, pharmacists, and customer. When a raw material provider delivers the raw material for the manufacturing purpose, he places a QR code on the raw material which includes all the details like location and timestamp. Once the raw material reaches the factor site, the manufacturer scans the QR code with the developed mobile application and verifies the details. Once the details are verified, the raw material can be used for the manufacturing process. When the manufacturing is done, the same QR code will be updated by the manufacturer with the additional details like manufactured date, expiry date, location, time stamp, and distributor name and batch number. The location can be updated with the mobile application because this mobile application cannot be run without the GPS permission of the mobile phone. With the GPS, the location can be embedded into the

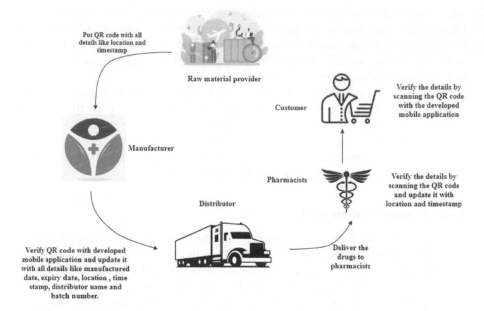

FIGURE 2.4
Drug supply chain based on blockchain.

QR code. Once the details are updated in the QR code, the manufactured drugs are handed over to the distributor for distribution. The distributor delivers the drugs to the pharmacists and he will check the authenticity of the drug by scanning the QR code with the developed mobile application. The pharmacists also update the QR code with the timestamp and the current location of drug delivery. Once the details are verified and updated, he will sell the drug to the customer. If a customer wants to verify the authenticity of the drug, he or she can verify the details by scanning the updated QR code with the mobile application. Once the details are verified, the customer can purchase the drugs from the pharmacists.

One of the major advantages of the proposed model is that whenever any entity involved in the supply chain tries to update the QR code, the location will automatically update because the mobile application that is used to update the details cannot run without enabling the GPS of the mobile phone. With the help of the proposed model, the inspection and the surveillance can be done much more efficiently because of transparency.

2.5 Research Challenges

Despite the fact that blockchain has immense possibilities to improve and add value to the medical sector, and that several companies have already begun testing it for particular use cases, there are a number of hurdles that must be overcome before its widespread adoption. The barriers to the widespread adoption of blockchain technology are multifaceted.

2.5.1 Integration and Compatibility with Older Systems

The healthcare space has a huge variety of technology, equipment, and components, not to mention employees, all of which work together to address the space's present demands. There are several opportunities for improvement, with blockchain technology promising to fill some of these gaps. However, blockchain technology would still be only one piece of the jigsaw, although a critical one. The blockchain technology must integrate effectively with current systems, and this will be a difficult task for a variety of reasons, including compatibility. The fact that the healthcare industry has a large number of devices and device kinds complicates matters even further. To make blockchain technology increasingly extensively utilised in medical, all necessary groups and partners must collaborate. If blockchain is unable to act as a catalyst for cooperation between users and responsible authorities, it is unlikely to have much value beyond being a simple and helpful solution for resolving some trust concerns. Many of the other challenges associated with ecosystem development will persist, independent of the blockchain. A significant component of this debate is the training of healthcare IT workers. If blockchain technology enters the healthcare field, these employees may need to be retrained on blockchain technology.

2.5.2 Uncertain Cost of Operation

Although blockchain has appealing features such as the lack of centralised authority, visibility, and relatively rapid payment resolution, the cost of building blockchain systems is unclear. Today, a substantial amount of healthcare resources are spent on employees, time, and money to create and operate the present conventional systems and data transfers. On top of that, there is the ongoing cost of upgrading systems, diagnosing problems, creating backups, and problem related to data leaks and hacks. Health Information Exchange (HIE) system based on blockchain might be more cost-effective and productive than the traditional HIE technique. This might be ascribed to enhanced safety by design, for instance. However, the entire cost issues connected with blockchain technology-based medical care must be thoroughly considered in the commercial and administrative model of a given medical centre.

2.5.3 Scaling

The underlying blockchain network must be scalable for any blockchain-based solution to be effective in any business, not just healthcare. Several healthcare solutions are expected to employ semi-permissioned blockchains, which are scalable and provides higher throughput for transactions at the expense of decentralisation, at least in the early stages. However, public blockchains will continue to be required for communication across permissioned blockchain networks. Furthermore, only highly scalable public blockchains can support a blockchain-enabled global HIE. In their current state, public blockchain networks such as Bitcoin and Ethereum are too slow and too expensive to host any decentralised applications on a wide scale.

2.5.4 Governance

The core idea of blockchain's distributed nature aids in bringing numerous parties into a trusted transaction system without the need for any centralised authorities. However, when it comes to how healthcare companies' function, there can be a variety of operational models. Some functional models of blockchain-based services may necessitate the involvement of a single stakeholder to operate as a regulator, overseeing the overall functioning of the blockchain. This management approach may be necessary to meet regulatory requirements, for example. It's uncertain how such a governance structure can be efficiently sustained in a network with so many distinct individuals. This aspect of governance will also be related to the suggested incentive schemes. However, as the usage of blockchain-based technologies in the healthcare business expands, a range of governance solutions will emerge.

2.5.5 Adoption and Incentives for Participation

As previously said, the use of blockchain technology in the healthcare sector will necessitate the coordination and cooperation of many parties. These stakeholders might include the hospital, gadget makers, medical professionals, patients, and so on. Since implementation of blockchain technology might require the collaboration of numerous actors, and some modifications to all of these partners' existing operational and management designs, it is normal for these interested parties to anticipate a few incentives to participate in the process of change. To meet these demands, new business models should be researched that provide equal rewards to all participants. The provided rewards must account for the costs and manpower associated with establishing or deploying blockchain-based technology and any fundamental changes to the operating framework that the deployed solution needs. Take, for example, the blockchain-based patient information monitoring platform. In this instance, a fair reward structure in the manner of a sustainable system that patients, device makers, backend IT solution providers, institutions, and others may accept must be developed. Since this technology is still in its early phases, strong incentive systems will be scarce for some time. Likewise, any risk associated with a certain incentive model must be properly examined and quantified to the maximum extent possible.

2.6 Conclusion

Blockchain technology has the ability to tackle a number of issues that are now affecting the healthcare business. It can facilitate innovative medical services as a faith facilitator, and as a reward mechanism, it can permit new revenue streams which may result in a new dynamic between different health professionals such as clinicians and clients. A patient care paradigm and a worldwide HIE, for instance, may be accomplished through blockchain-enabled decentralised faith and reward systems. Likewise, blockchain-based decentralised networks/services in healthcare may reduce vendor lock-in issues. In this chapter, the introduction to blockchain technology along with the background concepts of blockchain is discussed. The chapter provides in-depth details of the blockchain applications in the different domains of healthcare. The chapter also discusses application of blockchain in supply chain management. A model has been proposed for a new paradigm for blockchain-based supply chain management. The chapter concludes with the research challenges for blockchain in the healthcare domain.

References

Adamu, A. A., Salau. A. O., Zhiyong, L. (2021). A Robust Context and Role-Based Dynamic Access Control for Distributed Healthcare Information Systems. Internet of Things, Taylor and Francis, CRC Press, pp. 1–21. DOI: 10.1201/9781003140443-10

Ambrosus.io. (n.d.). Retrieved July 30, 2021, from https://ambrosus.io/.

Andreev, R., Andreev, R. A., Andreeva, P. A., Krotov, L. N., & Krotova, E. L. (2018). Review of blockchain technology: Types of blockchain and their application. *Intellekt. Sist. Proizv.*, 16(1), 11–14. doi: 10.22213/2410-9304-2018-1-11-14.

Azaria, A., Ekblaw, A., Vieira, T., & Lippman, A. (2016). MedRec: Using blockchain for medical data access and permission management. *Proceedings -2016 2nd International Conference on Open and Big Data, OBD 2016*, pp. 25–30. doi: 10.1109/OBD.2016.11.

Benchoufi, M., Porcher, R., & Ravaud, P. (2017). Blockchain protocols in clinical trials: Transparency and traceability of consent. *F1000Research, 6*. doi: 10.12688/F1000RESEARCH.10531.5.

Brandon, R. M., Podhorzer, M., & Pollak, T. H. (1991). Premiums without benefits: waste and inefficiency in the commercial health insurance industry. *International Journal of Health Services : Planning, Administration, Evaluation, 21*(2), 265–283. doi: 10.2190/H824-R263-YL47-WRQD.

Clauson, K. A., Breeden, E. A., Davidson, C., & Mackey, T. K. (2018). Leveraging blockchain technology to enhance supply chain management in healthcare: An exploration of challenges and opportunities in the health supply chain. *Blockchain in Healthcare Today, 1*(0). doi: 10.30953/bhty.v1.20.

CoinDesk (n.d.). Making sense of blockchain smart contracts.Retrieved July 30, 2021, from https://www.coindesk.com/making-sense-smart-contracts

Dubovitskaya, A., Xu, Z., Ryu, S., Schumacher, M., & Wang, F. (2017). Secure and trustable electronic medical records sharing using blockchain. *AMIA Annual Symposium Proceedings, 2017*, p. 650. Washington, DC, USA.

Dujak, D., & Sajter, D. (2019). *Blockchain Applications in Supply Chain*, pp. 21–46. doi: 10.1007/978-3-319-91668-2_2.

Huang, Y., Wu, J., & Long, C. (2018). Drugledger: A practical blockchain system for drug traceability and regulation. *2018 IEEE International Conference on Internet of Things (IThings) and IEEE Green Computing and Communications (GreenCom) and IEEE Cyber, Physical and Social Computing (CPSCom) and IEEE Smart Data (SmartData)*, pp. 1137–1144. doi: 10.1109/Cybermatics_2018.2018.00206.

Jamil, F., Hang, L., Kim, K. H., & Kim, D. H. (2019). A novel medical blockchain model for drug supply chain integrity management in a smart hospital. *Electronics, 8*(5), 505. doi: 10.3390/ELECTRONICS8050505.

Katal, A., Sethi, V., & Lamba, S. (2021). Blockchain consensus algorithms: Study and Challenges. In: Choudhury, T., Khanna, A., Toe, T. T., Khurana, M., & Nhu, N. G. (eds) *Blockchain Applications in IoT Ecosystem*, pp. 45–64. Springer, Cham. doi: 10.1007/978-3-030-65691-1_4.

Kuo, T.-T., Kim, H.-E., & Ohno-Machado, L. (2017). Blockchain distributed ledger technologies for biomedical and health care applications. *Journal of the American Medical Informatics Association : JAMIA, 24*(6), 1211–1220. doi: 10.1093/JAMIA/OCX068.

Liu, X., Barenji, A. V., Li, Z., Montreuil, B., & Huang, G. Q. (2021). Blockchain-based smart tracking and tracing platform for drug supply chain. *Computers & Industrial Engineering, 161*, 107669. doi: 10.1016/J.CIE.2021.107669.

Mandl, K. D., Szolovits, P., & Kohane, I. S. (2001). Public standards and patients' control: How to keep electronic medical records accessible but private. *BMJ : British Medical Journal, 322*(7281), 283. doi: 10.1136/BMJ.322.7281.283.

MediLedger. (n.d.). Blockchain solutions for Pharma companies. Retrieved July 30, 2021, from https://www.mediledger.com/

Mettler, M. (2016, November 18). Blockchain technology in healthcare: The revolution starts here. *2016 IEEE 18th International Conference on E-Health Networking, Applications and Services, Healthcom 2016*. doi: 10.1109/HealthCom.2016.7749510.

Modum. (n.d.). Retrieved July 30, 2021, from https://www.modum.io/solutions.

Nakamoto, S. (n.d.). Bitcoin: A peer-to-peer electronic cash system.

Narayanaswami, C., Nooyi, R., Govindaswamy, S., & Viswanathan, R. (2019). Blockchain anchored supply chain automation. *IBM Journal of Research and Development, 63*(2). doi: 10.1147/JRD.2019.2900655.

Nugent, T., Upton, D., & Cimpoesu, M. (2016). Improving data transparency in clinical trials using blockchain smart contracts. *F1000Research, 5*. doi: 10.12688/F1000RESEARCH.9756.1.

Plotnikov, V., & Kuznetsova, V. (2018). The prospects for the use of digital technology "blockchain" in the pharmaceutical market. *MATEC Web of Conferences, 193*, 02029. doi: 10.1051/MATECCONF/201819302029.

Salah, K., Rehman, M. H. U., Nizamuddin, N., & Al-Fuqaha, A. (2019). Blockchain for AI: Review and open research challenges. *IEEE Access, 7*, 10127–10149. doi: 10.1109/ACCESS.2018.2890507.

Salau A. O., Marriwala N., & Athaee M. (2021). Data security in wireless sensor networks: Attacks and countermeasures. Lecture Notes in Networks and Systems (vol. 140, pp. 173–186). Springer, Singapore. doi: 10.1007/978-981-15-7130-5_13

The Hindu BusinessLine. (n.d.). Neurogress to raise $44 m via blockchain-based crowdfunding. Retrieved July 17, 2021, from https://www.thehindubusinessline. com/companies/neurogress-to-raise-44-m-via-blockchain-based-crowdfunding/article22745078.ece.

Wang, S., Ouyang, L., Yuan, Y., Ni, X., Han, X., & Wang, F. Y. (2019). Blockchain-enabled smart contracts: Architecture, applications, and future trends. *IEEE Transactions on Systems, Man, and Cybernetics: Systems*, 49(11), 2266–2277. doi: 10.1109/TSMC.2019.2895123.

Wood, D. D. (2014). Ethereum: A secure decentralised generalised transaction ledger.

Xia, Q., Sifah, E. B., Asamoah, K. O., Gao, J., Du, X., & Guizani, M. (2017). MeDShare: Trust-less medical data sharing among cloud service providers via blockchain. *IEEE Access*, 5, 14757–14767. doi: 10.1109/ACCESS.2017.2730843.

3

Challenges and Benefits of Combining AI with Blockchain for Sustainable Environment

Ebenezer Leke Odekanle
First Technical University, Ibadan

Bamidele Sunday Fakinle and Olayomi Abiodun Falowo
Landmark University

Oludare Johnson Odejobi
Obafemi Awolowo University

CONTENTS

DOI: 10.1201/9781003193425-3

3.1 Introduction

Environmental sustainability described a continuous improved environment where the rate of utilization of renewable resources as well as pollution and non-renewable resources reduction are made to continue progressively and indefinitely (Morelli, 2011). It involves continuous maintenance of the natural resources in such a manner that would not jeopardize potentials of coming generations to meet their environmental needs. It addresses the need to ensure that while humans engage in several activities to meet their needs for shelter, food, water, etc., no damage is caused to the environment and no depletion of non-renewable resources. Environmental sustainability is the most important determinant for the survival of living organisms, bearing in mind that the quality of life of living organisms depends on the continuous improvement in the quality of the environment.

Two of the prominent issues capable of threatening sustainable environment are insufficient energy supply and environmental pollution. Globally, energy demand has escalated immensely due to the growing rate of population, industrialization and development. In any society, accessibility to constant and affordable energy has been described as the hallmarks of infrastructural development (Odekanle et al., 2020) because of its positive contribution to socio-economic development of such society. Although several nations depend on fossil fuel for their energy supply, some threats to sustainable environment such as global climate change, environmental degradation and other environmental issues are traceable to over-reliance on fossil fuel as a major source of energy (Kong et al., 2021). The dependence of anthropogenic activities such as industrialization and mining on non-renewable resources has been the issue of great concern for several decades due to the negative effects of these activities on the environment and the living organisms. The need for the use of natural resources, which are not only replenishable but also environmentally friendly, is germane to having a sustainable environment.

Similarly, environmental pollution (either air, land or water) from municipal, industrial and agricultural activities constitutes a barrier to the realization of environmental sustainability, considering its health impacts on both properties and living organism in an environment. Government, industries, organizations and individuals are persistently searching for technologies that would provide remedies to the menace for the betterment of the society.

From the aforementioned, urgent attention that will prevent further environmental damage but rather restore healthy environment is a necessity. For instance, in recent years, these environmental issues have generated a lot of discussion, public opinions, outrages and debates which have resulted in the rising interests in new approaches, innovation and advanced technology-based intelligent system such as Artificial Intelligence (AI) and blockchain that can provide robust solutions to environmental issues. It is considered that an effective environmental protection and sustainability depend not only on the quantity but also on the quality of the available data for qualitative decision. The situation becomes more complex considering the high quantum of the data originating from different sources which makes the data non-uniform. Advanced technologies are therefore required for monitoring and data analysis, data storage and retrieval, as well as monitoring and information communication – all of which are key to sustainable environment.

While both AI and blockchain have individual inherent shortcomings, many of these shortcoming and bottlenecks can be overcome by combining effectively the two technologies (Panarello et al., 2018). The goal of this chapter is to examine the benefits of synergizing AI and blockchain as in relation to renewable energy and environmental pollution prevention and/or control, for sustainable environment. Specifically, attempts are made to briefly give some background to the two technologies and to give overview of renewable energy and pollution prevention vis-à-vis sustainable environment. Benefits of combining AI and blockchain for sustainable environment with reference to renewable energy and pollution prevention/control are also examined. Some attempts are also made to identify challenges that pose threats to the combining effects of these two technologies.

3.2 AI and Blockchain Technologies: Overview

The present age has witnessed several transformational technologies that continually revolutionize human activities and the way we live. AI is another outstanding field which allows a machine to have cognitive functions to learn, infer and adapt based on data it collects. A recent prediction from market research indicates that AI market will grow up to 13 trillion US dollars by the year 2030 (Salah et al., 2019). AI also offers various services in the energy sector, such as data digitization, demand forecasting, resource management, energy storage facilitation, system failure prediction, and prevention (Kumari et al., 2020).

Similarly, blockchain, also known as distributed ledger technologies, was originally designed to enhance the transactions in a distributed manner (Andoni et al., 2019; Kumari et al., 2020). This technology system involves combination of decentralized consensus mechanism, distributed ledger and cryptographic security standard (Salau et al., 2021). As an emerging technology, it offers new opportunities for decentralized market designs (Mengelkamp et al., 2018). Although there exist obstacles in the merge of AI and blockchain, the integration of the two technologies will transform intelligent city network design to build environmental sustainability (Chen et al., 2021). The collaboration between the AI and blockchain technologies is projected to build a sustainable and smart society that can facilitate the quality of life, competitiveness as well as urban activity and service productivity of today and future generations in term of social, economic, cultural and environmental needs.

Various analytics-based technologies in conjunction with data mining, machine learning and deep learning have been used in the Internet of Things (IoT) structures to make technologies smarter, capable of predictions and classifications (Priya et al., 2020). A blockchain-based technology was designed and simulated to explore local energy market between 100 houses, based on the distributed information and communication technology by Mengelkamp et al. (2018). The implementation provided insight into the economic evaluation of blockchain-based energy market as well as its technical implication. The continuous consumption of energy generates a tremendous amount of data through the smart meters and advanced metering infrastructure in smart grid (SG) (Kumari et al., 2020). Using the collected data through IoT applications to the benefit of energy sectors will open up an era of convenience (Priya et al., 2020). The advancement in new technologies such as AI and blockchain has a good potential to address existing gap in energy supply and energy demand and transform the SG system. AI can provide estimation of energy consumption by analysing the data obtained from the smart meters. These processes bring economic advantage for the consumers of the energy cloud management system towards a demand-side flexible marketplace (Kumari et al., 2020).

In 2014 blockchain technology was first used to address issues in energy market. A decentralized digital currency was introduced for prosumers and consumers of locally produced renewable energy with equivalent value based on currency exchange market (Mihaylov et al., 2014). This currency was generated by injecting green energy into the SG. Transitioning into sustainable energy was a big challenge in the energy informatics, due to inability to integrate the increasing amount of renewable energy generated into major grids (Goebel et al., 2014). For instance, integration of renewable energy into a major SG needs a solution from energy suppliers since this energy may be generated from different sources depending on the dominant technologies in the area. Likewise, management of energy, energy security and data for efficient operations has to be strategically addressed on the energy consumers' side. Numerous issues associated with energy management can be finally solved in sustainable manner by applying AI combined with blockchain technologies. The techniques involving AI may offer services such as load management, energy load forecasting, classification of consumer, and blockchain analysis and can offer constant data and reliable mechanism for secure energy management (Salah et al., 2019; Kumari et al., 2020).

3.3 Pollution Prevention and Renewable Energy for Sustainable Environment

3.3.1 Pollution Prevention for Sustainable Environment

Sustainable environment is that which is able to meet its present environmental needs without endangering the potentials and liberty of the succeeding generations from meeting their own needs. Pollution prevention is one of the bedrocks and essential elements of sustainable environment. In recent years, there have been growing concerns about the rising menace of environmental pollution and its devastating effects on ecosystem. Pollution is the resulting effects of the continuous increase in population density, vehicular volumes, global heating, agricultural and industrial activities (Odekanle et al., 2017). Majority of environmental problems and their impacts could be linked to anthropogenic activities, especially in urban centres.

For instance, air and particulate emissions emanating from various sources such as vehicles, industrial and combustion operations cause interference with the natural ambient air quality and thus provoke unhealthy atmosphere. It has been reported that about 7,000,000 deaths are recorded annually from preventable air pollution (Gorobei, 2020). The effects of poor air quality are overwhelming, ranging from acidification, global warming, ozone depletion, damage to ecosystem and building to other various adverse health effects on humans and animals. A secured and healthy air is a precursor to an improved natural environment, since air covers a crucial location within other constituents of the biosphere. The complex protective ecological duty being performed by atmospheric air is enormous and it is thus considered as most significant impacts on humans. The process of breathing by human which continues and ever uninterrupted leads to inhalation of about 13,000 L of air per person/day (Gorobei, 2020), and during the process, human system reacts and responds to the slightest alteration in the natural composition of the inhaled air.

Similarly, waste disposal problems have been a major source of land and water pollution problem, especially in developing nations (Muhammad et al., 2013). In many of these countries, wastes are dumped indiscriminately by the road sides and in open places or pits without regards to health environmental health. Dumping of wastes in streams, river channels and drainages is also common practice in some communities, and these actions are connected to continuous increase in flood and pollution of groundwater being experienced in such communities (Nkwachukwu et al., 2010).

Pollution prevention is vital to sustainable environment. It reduces the quantity of pollutants produced through a process having a design that increases the efficiency of a process and ensures judicious utilization of resources (Sherman et al., 2016; Elleuch et al., 2018). While pollution control strategy (*otherwise called end-of-pipe*) addresses, manages or treats pollutants or wastes after being generated in order to meet environmental guidelines (Petraru & Gavrilescu, 2010), pollution prevention (also known as cleaner technology) reduces the volume of pollutant before it is generated from the source. This approach is considered as a pro-active management of pollution from the source, leading to eco-efficiency and cleaner production. The principle of pollution prevention considers that wastes generation can be minimized, if not totally eliminated by increasing the efficient use of raw materials, natural resources and energy.

Pollution prevention is therefore an essential component of sustainable environment because the best way to create a clean and habitable environment is to reduce pollution generation. This can be achieved by designing an environmentally friendly and smart process through a technology-based intelligent system that understands how wastes are generated, minimized and/or prevented as well as how raw material, natural resources and energy can be efficiently utilized. Pollution prevention will reduce the waste disposal problems and cost, create healthier environments and enhance sound environmental management. Therefore, the idea of avoidance of pollutants generation, rather than end-of-pipe treatment, should be the first priority in an environmental management hierarchy among which are prevention, treatment, recycling, etc.

3.3.2 Renewable Energy for Sustainable Environment

Energy sources in the modern energy distribution system show that the demand for energy continues to increase at a huge pace due to large-scale IoT devices, other electric device and improvement in cyber-physical system (Kumari et al., 2020). Also, industrialization and population increase across the world are other factors responsible for increase in energy demand (Falowo et al., 2019). The conventional energy sources that serve the existing consumers are finite in nature, apart from accompany environmental pollutions in its usage. Considering the economic growth and prosperity of many advanced countries since the beginning of industrial revolution, energy plays a vital role, if not the most important among others. More so, this internet age makes the use of sustainable energy paramount, in order to optimize several developments discovered in preserving the environment. Sources of conventional energy such as coal, fossil fuel, nuclear and thermal power plant, due to their characteristics, cannot reasonably sustain the growing energy demand. For this reason, renewable energy sources such as hydro, wind, solar and biomass can fulfil a significant role in addition to conventional energy in generating energy that will meet consumers demand. Using renewable sources of energy for large-scale generation of energy instead of the conventional energy resources will enhance better maintenance of ecosystem (Bali & Kumar, 2016). Renewable energy is now considered as the most economically and environmentally friendly alternative energy, especially based on the cost and creativity involved (Chen et al., 2021). While this assertion is true in advanced world, developing nations are far off due to the level of their technology. The renewable energy resources particularly in developing countries are persistently underutilized, and the amalgamation of renewable energy technologies into an existing SG or as self-sustaining application requires an accurate assessment of renewable energy sources. This information is not readily available in many countries or regions particularly in developing nations. The role of renewable energy technology in small islands developing state on electricity consumption has been reviewed (Weisser, 2004). As it stands, renewable energy sources usage has potential to undercut the cost of generating electricity. For developing countries, this remains a big challenge due to lack of technological advancement in adopting suitable energy technology that will improve their economy. The compatibility of multiple energy sources in most cases helps to maximize energy consumption (Chen et al., 2021).

With all the hype surrounding utilization of renewable energy sources, there remain persistent problems due to its instability. For this reason, microgrids are considered a strategic solution to efficiently manage and maintain renewable energy generation. In comparison with fossil fuel, this technology is still lacking behind due to its price. The technology must of course be improved in order to bring down cost prices associated with wind and solar energy sources.

The management of diverse energy generation from both conventional and non-conventional sources at the supply side and consumers receiving the energy distribution on the demand side remains a big challenge. Particularly on the supply side of energy, harnessing this energy from different sources into a single supply line is constrained by limited SG infrastructure. SG is a complex system which is being touted as the next generation of electrical power system (Melhem et al., 2016). Another option is to manage the energy at the demand side using the renewable energy sources known as energy cloud. This service-based technology will manage intelligently energy consumption and offer electricity as well as other value-added service to consumer on demand. An advanced renewable energy distributed generation will provide reliable and adequate power and energy considering the dynamic existence of renewable energy supplies, with proper balance and regulation (Chen et al., 2021). To have sustainable energy, the input of policy makers and external economic stresses on the cost of electricity generation is essential to model accurately cost-effective renewable energy. In order to achieve this, sustainable model of energy generation requires different areas of research to highlight the key parameters and their interactions in a way that positions renewable energy system for future takeover.

3.4 Benefits of Combining AI with Blockchain for Environmental Sustainability

Several peculiar bottlenecks in AI and blockchain can be addressed by integrating the two technologies. Some of the benefits are as follows.

3.4.1 Energy Cost Reduction

The generation and distribution of energy from the energy supplier to the consumer involves several sub-networks and facilities managed by central unit. Each facilities and energy sub-network is handled by different professionals which eventually increase the cost of generating electricity. AI combined with blockchains could help address the challenges faced by decentralized energy systems. Due to the integrity of blockchain technology in securing end-to-end transaction particularly, the task that involves complex components, AI combined with blockchain, can eliminate the need for central management by automating the technology and subsequently reducing energy generated cost.

3.4.2 Security of Stored Environmental Data and Information for Valid Decisions and Policies

AI algorithms depend on the information to make decision. These algorithms perform better when the data are obtained from a secure, reliable and credible origin. This aspect of security and credibility of data storage can be provided by blockchain. Blockchain data are stored with high level of security and integrity which cannot be easily altered. Synergizing AI and blockchain can lead to having a credible, secure and trusted system for sensitive data storage which AI can access, learn and utilize to make environmental decisions and policies. The outcomes of such decision are undisputable and credible.

Generally, dysfunction of decisions occurs when environmental decisions that originated from AI platform become unreliable and difficult to understand. However, since blockchain transacts data in a distributed and decentralized ledger, this attribute will assist environmental protection agencies to make policies and decisions that will gain public acceptance and trust, bearing in mind with high certainty that the records have not been compromised especially during the process of auditing. Technology-based intelligent system that will produce valid decisions on environmental management must include qualitative analytical component which must not only be user-friendly but also integrate a secure, transparent and reliable component that will enhance valid and trusted decisions. In other words, because the stored environmental data are secure and tamper-proof, transparency of the decision support system provided by AI/blockchain combination will improve.

3.4.3 Global Cooperation in the Fight against Climate Change

Climate change is one of the major environmental issues being faced by several nations of the world and thus requires a technology-based intelligent approach for solution. Consolidation of AI with blockchain can create an opportunity to combat menace of climate change through its transparent nature that will enhance greater stakeholders' participation, trust and innovative solution to produce improved climatic action. Combination of AI and blockchain will produce, build and manage a decentralized structure and globally accepted solutions to climate change. This approach will result in stronger cooperation among multiple international stakeholders and help in policy formulations that focus on compliance to CO_2 emission, carbon-offset credit and overall climate change mitigation. Combination of AI and blockchain can improve climate forecast and accuracy of prediction. The more the climate changes, the more the accuracy of the forecast is required. Since climatic models usually result in different forecast probably because of the nature of the data and the way such data are processed. However, with the combination of AI and blockchain, several complex dynamic parameters can be incorporated into the model calculation to give accurate prediction model which can help decision makers to make more useful forecast.

3.4.4 Rapid Response to Environmental Emergency and Prevention of Environmental Hazards

Another benefit of combining AI with blockchain is seen in its capability, not only to assist environmental emergency response team to response to emergency but also to prevent environmental hazard. The multidisciplinary and complex nature of environmental problems may result into hazards or accidents. However, identification and development of qualitative and quantitative analysis of such hazards can be made easy and more valid by the combining effects of decentralized AI (AI/blockchain combination). For instance, trusted and reliable information from decentralized AI can be utilized to create a technology-based intelligent system that will guide emergency response team on how to tackle industrial or environmental hazards (Avouris, 1995). Since heterogeneity/non-uniformity of the data obtained from different sources have been treated and overcome by the synergy between AI and blockchain, identification, evaluation and assessment of risk and degree of potential hazard can be numerically or qualitatively modelled. Hence, prediction of potential environmental hazard can be made with such degree of accuracy, and this action can assist government agency that is responsible for environmental protection as well as response team to develop adequate mitigating or preventive measures.

3.4.5 Less Energy Intermediaries

AI combined with blockchain technology will remove intermediaries, thereby creating a new paradigm in management of energy. Generation, transfer and distribution of energy especially in a country having multi-energy sources will undoubtedly require large intermediaries. This includes energy suppliers, utility companies, network operators, aggregators and generators (Andoni et al., 2019). AI combined with blockchain can facilitate peer-to-peer energy trading without involving usual intermediaries.

3.4.6 Energy Digitization

Blockchains can be viewed as data banks that allow numerous users to make alterations in the ledger concurrently, which can result in various chain versions. The implication of this is that energy consumers can be served with different energy suppliers within an integrated AI-blockchain technology. Blockchain alongside AI will facilitate data collection, analysis, storage and management from different sectors involved in sustainable energy. Since large data available for energy companies are still process manually thereby creating gaps which prevent efficient implementation of energy produced from different sources. AI integrated with blockchain can innovatively produce cost-efficient technology capable of making sustainable energy competitive.

3.4.7 Serve as Tool for Air Pollution Monitoring for Cleaner Environment

Environmental monitoring plays a vital role in evaluating the health and safety challenges in our environment. Increase in anthropogenic air emission globally continues to result in decrease in the ambient air quality of our environments. It has been reported that nine out every ten people inhale polluted air on a daily basis and about 7 million deaths have been attributed to air pollution (World Health Organization, WHO, 2016; Gorobei, 2020). Several cities, especially in developing nations, cannot meet globally accepted air quality guidelines; hence there is prevalence of various health challenges such as respiratory diseases which include asthma and lung cancer (WHO, 2018; Odekanle et al., 2020). Continuous improvement of the quality of our environment can be achieved through the use of technologically driven intelligent, secure decision making and tamper-proof air emission monitoring systems which will be both publicly and privately available for users – these can be offered by AI/blockchain combination. Air pollution monitoring systems generally collect data in the air and categorize the data qualitatively to reflect the status of the air quality. The air pollution monitoring systems are expected also to collect data from various sources, validate the data and make the data as accessible as possible. Apart from these, the data should be robust, secure and tamper-proof; all these will assist the users, either independent researchers or government agency to analyse the data with integrity, certainty and confidence to produce valid outcomes or decisions.

AI is centralized and this creates the possibility or loopholes for environmental data to suffer being tampered with. This then compromises the authenticity of the data. On the other hand, blockchain is decentralized in nature. Its synergy with AI will provide higher immutability and prevent the possibility of tampering with the air quality data. Apart from the fact that information stored within blockchain is secure, it also has high capability for sharing sensitive and unalterable information. Therefore, combination of AI and blockchain for air monitoring will enhance credibility and reliability of the decision/formulated policy for cleaner environments. The synergy can also be an effective and efficient tool for the identification of sources of air emission of air pollutants. Sources of leakages or potential leakages in pipe can be identified accurately and quickly, air quality can be monitored easily from one location to the other, fuel efficiency in automobiles can be enabled, and volumes of traffic on major roads can be monitored to adjust the flow of traffic, thereby reducing vehicular emissions. Air pollution can be modelled and predicted more easily and accurately, based on the integrity and reliability of the information obtained from the synergy of these two technologies.

3.4.8 Pollution Reduction

High quantum of waste is generated on daily basis without much concern for its disposal. This practice continues to grow worse in the developing nations possibly because of the level of poverty and exponential population growth (Atta et al., 2016). These wastes do not only constitute a menace to effective environmental management but are also associated with potential antimicrobial resistance pattern (Fearson et al., 2014). Combining AI with blockchain could prove beneficial and be an effective way of alleviating the problem of wastes disposal. For instance, e-waste generation can be reduced by refurbishing and re-using unwanted electronic materials through the use a more robust technology-based intelligent system obtained from the synergy of AI and blockchain. Life cycle of material can easily be detected and the consumer notified and guided on how to get it replaced. This innovative system can also assist in the conversion of organic wastes to other useful products. Apart from the pollution control which involves management or disposal of wastes after generation, AI/blockchain combination can provide an innovative platform for waste prevention, where wastes minimization/prevention is factored into manufacturing process design or route by designing a process that maximizes efficient use of raw materials.

3.4.9 Enhances Accountability and Transparency in the Delivery of Humanitarian Relief Packages

With the current situations in our environments, ranging from natural disaster to several other challenges brought about by untold economic hardship and terrorism, the need for humanitarian aids cannot be under-estimated. For the humanitarian organization to be effective and efficient in the discharge of its duty of delivering aid, it is paramount to take the advantage of the emerging technologies and innovation. Combination of AI with blockchain is considered an effective tool which can transform humanitarian organizations and their service delivery by enhancing transparency and accountability of both the donors and the recipients of the aids. Flow of information can be easily traced, cost can be reduced, records of location where the humanitarian aid is required and eventually directed can be monitored, and corruption can, at large, be prevented. Also, since in the blockchain, the security of the data or information kept in the distributed ledger is cryptographically maintained, the information is visible to the public and private to the user. This implies that the data are unchangeable – though available for all. This process will encourage auditability, better and easy reconciliation, and therefore humanitarian organization gain the trust of both the communities and the donors.

3.4.10 Reduction of Carbon Footprint

A major challenge in the developmental approach to low industrial carbon emission is the pathway or route to low carbon energy system of an industry (Rui et al., 2016). With the combination of AI and blockchain, emission of carbon can be reduced, thereby reducing emission of greenhouse gases both in transport and other sectors. This could be achieved by gathering data from the sensors located in the pilot plants at the industries. The data collected from different sources are securely stored on blockchain platforms and are later analysed and processed using AI algorithm to make informed decision on how to place premium on process efficiency as well as material and energy savings. AI/blockchain integrated platform can help to evaluate carbon footprint as well as reporting carbon emission trading – all of which will assist in the prediction of the next line of action (Osuji, 2020).

3.4.11 Increased Energy Security

Blockchain system integrated with AI technology can conduct energy trading in an efficient manner since it is highly secure and tamper resistant (Andoni et al., 2019). AI algorithm can handle heterogeneous data whether structure or unstructured, forecast the consumer energy demand and recommend exact energy in order to avoid energy wastage (Kumari et al., 2020). The added security indicates that the records cannot be easily manipulated because they are encrypted by design. The integration of these two technologies provides the ease of trading energy by different energy supplier. A consumer can therefore receive energy supply not only from a single trusted centre but from different energy provider. Each renewable energy source provider can supply a single consumer, and a copy of the succession of record can be held by individual energy supplier, apart from the fact that the agreement on the state of ledger can be reached harmoniously. Using blockchain, activities are connected to the preceding ones by cryptography that enhances resilience and high security of the blockchain network. Individual users can assess the validity of the activities, thereby making the whole record transparent and tamper-proof (Salah et al., 2019; Kumari et al., 2020).

3.4.12 Decision and Energy Traceability

The combination of AI and blockchain technology will enhance energy sustainability by making informed decision based on the data collected to manage energy. Decision made by AI using energy data on energy cloud management can be traced at every chain from energy generator to the consumer by all participating members through blockchain technology. The information on energy distribution as well as decision making by the technology can easily be assessed by the participating members.

3.5 Challenges of Combining AI with Blockchain

Combining AI and blockchain is beneficial to sustainable environment by providing an innovative technology that can recreate and restructure our environment. However, no technology is entirely protected against challenges. For the synergy of AI and blockchain to deliver the potential benefits for sustainable environment, there is need to address several challenges of the synergy. Although the challenges are regulatory and technical, they are surmountable. The following are some of the challenges that could be encountered by combining AI and blockchain.

3.5.1 Data Security

Although blockchain is known for its provision of secure and tamper-proof information, this technology is also prone to cyber-attacks (Li et al., 2017; Kumari et al., 2020). The decentralized or distributed ledger in blockchain can be attacked and exposed to abuses, thereby making the decentralized mechanism becomes centralized within few users. These attacks therefore compromise the tamper-proof nature and quality of the data stored in the blockchain. When such compromised data are fed into AI application, the decisions or output obtained are biased and do not reflect the true nature of the situation being analysed. Models, forecasts and prediction from such data are thus unreliable and will not provide the needed solutions to the menace intended to be addressed.

3.5.2 Issue of Standardization, Regulation and Governance

Currently, blockchain technology is not yet standardized and regulated (Gulati et al., 2020). There is a lot of ongoing processes by several regulatory bodies to establish governing standards and regulation that will ensure interoperability, integration and governance (Anjuma et al., 2017). For effective deployment of AI in blockchain application, it is expected that guidelines, policies and regulation must be established for effective coordination. Also, the governance of the platform could constitute another bottleneck because management and deployment of blockchain platform among several users could be a difficult job. There could also be some challenges relating to the type of blockchain platform to deploy, who manages the platform and at what location must it be deployed as well as the regulation to follow. Methodology of how consensus will be reached among different energy suppliers is a continuing research area and might vary to suit different range of energy domains (Andoni et al., 2019). Therefore, there should be sets of standards and governing guidelines that would ease the application of blockchain in AI.

3.5.3 Quality of Data

The main function of blockchain is to store data in a secure and tamper-proof manner and to transfer such data for subsequent use. AI on its own collects data, trains and processes the data to make decisions. AI therefore depends mainly on the data transferred into it. Its combination with blockchain implies that the stored data can be transferred to it to make decisions based on the data collected. However, the quality of the data received by AI determines the quality of the output generated by AI and thus its proper application. Feeding wrong data into AI platform will produce a wrong decision and vice versa. Also, when AI goes wrong or make a biased decision especially based on the wrong data set fed into it, it would be very difficult for user to detect when this happens and what to do when such mistake occurs. Similarly, to work effectively, AI application requires large quantity of data; otherwise, the analysis of the problem may be difficult (Gulati et al., 2020). Collection or gathering of the required high quality and large quantity of data may be problematic.

3.5.4 Lack of Professional Expertise

Lack of competent talent is another major challenge in combining AI and blockchain. Competent professional to develop or formulate algorithm that work in a decentralized or distributed manner is a requirement for effective application of blockchain technology. Professionals in the field, who really understand the workability of the integration of the technology, are still lacking, and this could be a drawback to the benefits that are meant to be derived from the combination of the two technologies.

3.5.5 Privacy

For the combination of AI and blockchain, it is fully known that blockchain ledger supplies tamper-proof and secure data to AI platform, the data also are made available to public users. However, blockchain platform could be invaded and manipulated by hackers. This could constitute a significant threat to the security of the data. Although private blockchains are considered relatively safer than the public blockchain platform, private blockchain ledger has a limited capacity for storing huge amount of data required by AI to perform properly and maximally.

3.5.6 Scalability and Side Chains

One other major issue facing blockchain technology is its scalability. For blockchain platform, its performance is extremely low compared with other platforms like Facebook (Salah et al., 2019) which deals with several millions of data or transaction every second. Although side channels or side chains could be employed to increase the performance of blockchain, more research is still required to increase the scalability of blockchain.

3.5.7 Inaccessibility to Digital Infrastructure

Many rural communities in developing nations have no access to digital infrastructure, and this could constitute a major drawback in accessing the full benefits of the synergy of the two technologies. In many of these countries, the rural centres who are the custodians of the natural resources have no access to digital facilities, apart from the prevalence of lack of basic understanding of the workability of digital system. For AI and blockchain combination to provide its full deliverables, there must be an even playing-ground among all the participants, in terms of availability of infrastructure, regardless of location (rural or urban). Previous report revealed that, as at 2016, about 82% of people in developed countries get access to internet, as against 14% in developing nations (World Bank, 2018). Access to digital facilities will enhance digital literacy among various participants and stakeholders, boosting their understanding of the relevant basic concepts (Gatteschi et al., 2018).

3.6 Conclusion and Recommendation

3.6.1 Conclusion

Combination of AI with blockchain is known as decentralized AI. The two technologies have individual abilities to assist in sustainable environment; however, this chapter focused on the benefits and challenges of the combination of the two technologies, especially as it relates to sustainable environment vis-à-vis renewable energy and pollution prevention. In this chapter, brief introduction and background information about AI and blockchain were given; renewable energy and pollution prevention as means for sustainable environment were also discussed. The last sections were devoted to the benefits and challenges of combining AI with blockchain. As discussed, integrating AI with blockchain can enhance global cooperation in the fight against climate change, assist in rapid response to environmental emergency and prevention of environmental hazards and help in the effort to reduce pollution. Apart from these, the synergy can serve as effective advanced air pollution monitoring tool, reduce carbon footprint and can enhance effective delivery of humanitarian relief aids.

To be able to deliver all the potential benefits that could be offered by the synergy of the two technologies, a number of challenges must be overcome. Issues relating to governance, reaching consensus among major players (e.g. energy suppliers/environmental protection agency) not yet fine-tuned, privacy and scalability of the platforms are serious bottlenecks for the combination. Also, lack of standardization for blockchain, lack of professional expertise in integrating the technology, high development cost, security vulnerabilities and software bugs due to cyber-attack, insecure data analytics due to misleading data as well as access to digital infrastructures are serious challenges militating against combining AI and blockchain technologies. It is therefore concluded that combining AI and blockchain can be a good approach towards sustainable environment, despite the surmountable challenges.

3.6.2 Recommendations

To overcome some of the challenges of combining AI and blockchain, and to realize more of the benefits of the amalgamation the two technologies, the following recommendations or possible solutions are suggested:

- More advanced research work needs to be carried out to address shortcomings relating to security and privacy of data in blockchain technology.
- Both government and industries need to support innovative ideas tailored towards AI and blockchain solutions as well as the integration of the two platforms.
- Governments, especially in developing nations, need to invest in innovations and digital facilities as well as digital literacy to increase understanding and smooth utilization of the technologies, even in rural communities.
- Capacity building and development of expertise and relevant professionals that will shape the potential opportunities created by AI and blockchain combination is also advocated.

References

Andoni, M., Robu, V., Flynn, D., Abram, S., Geach, D., Jenkins, D., McCallum, P. and Peacock, A. (2019). Blockchain technology in the energy sector: A systematic review of challenges and opportunities. *Renewable and Sustainable Energy Reviews*, 100: 143–174.

Anjum, A., Sporny, M., and Sill, A. (2017). Blockchain standards for compliance and trust. *IEEE Cloud Computing*, 4: 84–90.

Atta, A.Y., Aminu, M., Yususf, N., Gano, Z.S., Ahmed, O.U., and Fasanya, O.O. (2016). Potential of waste to enrgy in Nigeria. *Journal of Applied Sciences Research*, 12: 1–6.

Avouris, N.M. (1995). Cooperating knowledge-based system for environmental decision support. *Knowledge-Based System*, 8: 35–53.

Bali, R.S. and Kumar, N. (2016). Secure clustering for efficient data dissemination in vehicular cyber–physical systems. *Future Generation Computer Systems*, 56: 476–492.

Chen, C., Hu, Y., Marimuthu, K., and Kumar, P.M. (2021). Artificial intelligence on economic evaluation of energy efficiency and renewable energy technologies. *Sustainable Energy Technologies and Assessments*, 47: 101358.

Elleuch, B., Bouhamed, F., Elloussaief, M., Jaghbir, M. (2018). Environmental sustainability and pollution prevention. *Environemntal Science Pollution Research*, 25: 18223–18225.

Falowo, O.A., Oloko-Oba, M.I., and Betiku, E. (2019). Biodiesel production intensification via microwave irradiation-assisted transesterification of oil blend using nanoparticles from elephant-ear tree pod husk as a base heterogeneous catalyst. *Chemical Engineering and Processing-Process Intensification*, 140: 157–170.

Fearson, J., Mensah, S.B., and Boateng, V. (2014). Abattoir operation, waste generation and management in the Tamale metropolis: Case study of the Tamale slaughterhouse. *Journal of Public Health and Epidemiology*, 6: 14–19.

Gatteschi, V., Lamberti, F., Demartini, C., Prantenda, C., and Santamaria, V. (2018). Blockchain and smart contracts for insurance: Is the technology mature enough? *Future Internet*, 10(2): 20.

Goebel, C., Jacobsen, H.-A., del Razo, V., Doblander, C., Rivera, J., Ilg, J., Flath, C., Schmeck, H., Weinhardt, C., and Pathmaperuma, D. (2014). Energy informatics. *Business & Information Systems Engineering*, 6: 25–31.

Gorobei, M, (2020). Environmental sustainability and pollution prevention: The negative impact of carbon-containing dust on the environment and humans and effective measures for its reducing. *International Journal of Advanced Research*, 8: 1489–1496.

Gulati, P., Sharma, A., Bhasin, K., and Azad, C. (2020). Approaches of blockchain with AI: Challenges and future direction. *Proceedings of the International Conference on Innovative Computing & Communications (ICICC) 2020*, Available at SSRN: https://ssrn.com/abstract=3600735 or http://dx.doi.org/10.2139/ssrn.3600735.

Kong, K.G., How, B.S.H., Teng, S.H. et al., (2021). Towards data-driven process integration for renewable energy planning. *Current Opinion in Chemical Engineering*, 31: 100665.

Kumari, A., Gupta, R., Tanwar, S., and Kumar, N. (2020). Blockchain and Ai amalgamation for energy cloud management: Challenges, solutions, and future directions. *Journal of Parallel and Distributed Computing*, 143: 148–166.

Li, X., Jiang, P., Chen, T., Luo, X., and Wen, Q. (2017). A survey on the security of blockchain systems. *Future Generation Computer Systems*, doi: 10.1016/j.future.2017.08.020.

Melhem, F.Y., Moubayed, N., and Grunder, O. (2016). Residential energy management in smart grid considering renewable energy sources and vehicle-to-grid integration. *Paper Presented at the 2016 IEEE Electrical Power and Energy Conference (EPEC)*.

Mengelkamp, E., Notheisen, B., Beer, C., Dauer, D., and Weinhardt, C. (2018). A blockchain-based smart grid: Towards sustainable local energy markets. *Computer Science-Research and Development*, 33: 207–214.

Mihaylov, M., Jurado, S., Avellana, N., Van Moffaert, K., de Abril, I.M. and Nowé, A. (2014). Nrgcoin: Virtual currency for trading of renewable energy in smart grids. *Paper Presented at the 11th International Conference on the European Energy Market (EEM14)*. Cracow (Kraków), Poland.

Morelli, J. (2011). Environmental sustainability: A definition for environmental professionals. *Journal for Environmental Sustainability*, 1: 1–9. doi: 10.1448/jes.01.0002.

Muhammad, M.U., Abubakar, I., Musa, A., and Kamba, A.S. (2013). Utilization of waste as an alternative energy sources for sustainable development: A review. *ChemSearch Journal,* 4: 57–61.

Nkwachukwu, O.I., Chidi, N.I., and Charles, K.O. (2010). Issue of roadside disposal habit municipal solid waste environmental impacts and implementation of sound management practices in developing country: Nigeria. *International Journal of Environmental Science of Science and Development,* 1: 409–418.

Odekanle, E.L., Fakinle, B.S., Jimoda, L.A., Okedere, O.B., Akeredolu, F.A., Sonibare, J.A. (2017). In-vehicle and pedestrian exposure to carbon monoxide and volatile organic compounds in a mega city. *Urban Climate,* 21: 173–182.

Odekanle, E.L., Sonibare, O.O., Odejobi, O.J., Fakinle, B.S., and Akeredolu, F.A. (2020). Air emissions and health risk assessment around abattoir facility. *Heliyon,* 6: e04365. doi: 10.1016/j.heliyon.2020.e04365.

Osuji, N. (2020). Blockchain technology for monitoring and reporting of carbon emission trading: A case study and its possible implementation in the danish energy industry. Master Thesis. Technical Faculty of IT and Design, Aalborg University in Copenhagen.

Panarello, A., Tapas, N., Merlino, G., Longo, F., and Puliafito, A. (2018). Blockchain and IoT integration: A systematic survey. *Sensors,* 18(8): E2575.

Petraru, M., and Gavrilescu, M. (2010). Pollution prevention, a key to economic and environmental sustainability. *Environmental Engineering and Management Journal,* 9: 597–614.

Priya, R., Bhattacharya, S., Maddikunta, P.K.R., Somayaji, S.R.K., Lakshmanna, K., Kaluri, R., Hussien, A., and Gadekallu, T.R. (2020). Load balancing of energy cloud using wind driven and firefly algorithms in internet of everything. *Journal of Parallel and Distributed Computing,* 142: 16–26.

Riu, Z., Ning, M., Yong, G., and Yulong, H. (2016). Allocation of carbon emission among industries/sectors: An emission intensity and reduction constrained approach. *Journal of Cleaner Production.* doi: 10.1016/j.jclepro.2016.10.159.

Salau A. O., Marriwala N., and Athaee M. (2021). Data security in wireless sensor networks: Attacks and countermeasures. *Lecture Notes in Networks and Systems,* Vol. 140, pp. 173–186. Springer, Singapore. DOI: 10.1007/978-981-15-7130-5_13

Salah, K., Rehman, M.H.U., Nizamuddin, N., and Al-Fuqaha, A. (2019). Blockchain for AI: Review and open research challenges. *IEEE Access,* 7: 10127–10149.

Sherman, J.M., McGain, F., FANZCA, and FCICM (2016). Environmental sustainability in anaesthesia pollution prevention and patient safety. *Advances in Anaesthesia,* 34: 47–61.

Weisser, D. (2004). On the economics of electricity consumption in small Island developing states: A role for renewable energy technologies? *Energy Policy,* 32: 127–140.

WHO (2016). Ambient air pollution: a global assessment of exposure and burden of disease (2016). https://www.who.int/phe/publications/air-pollution-global-assessment/en/. Accessed on June 8, 2021.

WHO (2018). Air quality and health. http://www.who.int/mediacentre/factsheets/fs313/en/. Accessed on June 10, 2021.

World Bank (2018). Individual using internet. World Bank data. https://data.world-bank.org. Accessed on June 28, 2021.

4

Chain of Transactions: How Blockchain Is Changing the Landscape of FinTech

Suhasini Verma
Manipal University Jaipur

Andrea Milagros Carrasco Suyo
University of Piura

Jeevesh Sharma
Manipal University Jaipur

CONTENTS

DOI: 10.1201/9781003193425-4

4.1 Introduction

The fast pace of development of technology has disrupted almost all sphere of life, and businesses are no exception. New software, databases and systems have been launched for optimizing processes and help businesses to have a more agile management. Because of the current circumstances and technology development, more and more enterprises are entering into the digital world, and the disruption is so apparent that a new term FinTech is extensively coined since past decade to distinguish the extensive use of technology in the world of finance.

4.2 Literature Survey

For the banking and financial services industries, blockchain technologies provide a number of appealing features. This system is durable and may function as decentralized networks without the need of a central server. Organizations employ decentralized open-source standards to operate (Treleaven et al., 2017). Blockchain technology can enhance corporate operations in the banking and financial services industry while also establishing secure, reliable records of agreements and transactions. The study elaborated by Fosso Wamba et al. (2020), titled "Bitcoin, blockchain and FinTech: a systematic review and case studies in the supply chain", studied and analysed 141 articles and elaborated the applications of blockchain technology in supply chain management sector. The paper concluded that these technologies are in constant development and evolution, and its use can contribute to generate a competitive advantage for businesses. Intermediaries frequently assist transacting parties in finding each other, establishing confidence, and settling transactions in economic transactions (Roth, 2015). Recent advancements in blockchain technology have paved the way for a new paradigm based on decentralization and disintermediation. Through dispersed trust and decentralized platforms, blockchain technology can remove the necessity of middlemen in financial transactions and allows peer-to-peer transactions (Chen & Bellavitis, 2019).

Cryptocurrency is among the most significant application of the blockchain technology. Since the launch of the first cryptocurrency, Bitcoin, in 2009, several other cryptocurrencies have emerged to meet a variety of demands and goals. Furthermore, as this technology has eliminated the role of intermediary, the time and financial cost have reduced significantly. There have been attempts to apply blockchain in a variety of areas, including banking and manufacturing, and its usage in the medical field is also being investigated (Hashemi et al., 2019). The COVID-19 pandemic, which resulted in significant mortality and disruption, exposed flaws in current systems that protect human health and well-being. Inadequate and delayed data, as well as widespread misinformation, are generating significant harm and increasing the confrontation between data privacy and public health issues. The distributed governing system and confidentiality features of blockchain can be used to create a system that can help resolve the tension between maintaining privacy and meeting public health needs in the fight against COVID-19 (Khurshid, 2020). To alter the data stored on a blockchain, it is required to concurrently edit the data divided among the participants. This makes falsifying or manipulating data very difficult, as well as ensuring its trustworthiness and openness. When individuals with infectious diseases are identified, diagnostic information and clinical presentation may be shared swiftly and precisely (Chang & Park, 2020).

4.3 FinTech

FinTech can broadly be defined as "use of technology to deliver financial solutions" (Arner et al., 2016). FinTech encompasses three dimensions: Input (organization, money, and technology) mechanism (applied IT to finance, disruption, etc.) and output (new products/processes/business models) (Gai et al., 2018, Philippon, 2016, Goldstein et al., 2019, and Zavolokina et al., 2016). Blockchain, Cryptography, Internet of Things (IoT), Artificial Intelligence (AI), Machine Learning (ML), etc. are few of the technologies, which are being extensively used on finance segment to provide services like payment, lending, investments, and even currency. The use of technology makes the processes faster, easier, trustworthy, and speedy as no intermediary intervention is required (see Figure 4.1). According to EY (2019), FinTech adoption worldwide has increased 60%, between the years 2015 and 2019. FinTech offered many a service for the perusal of its clients as depicted in Figure 4.2.

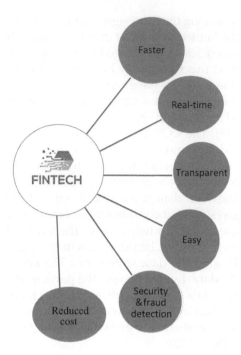

FIGURE 4.1
Advantages that FinTech can offer to people and enterprises. (Fosso Wamba et al., (2020).)

FIGURE 4.2
Principal examples of services offered by FinTech. EY (2020).

4.3.1 The Main Differences between FinTech and Traditional Finance

Following are the main differences between FinTech and traditional finance:

a. Traditional finance commonly has intermediaries in between, which increase transaction cost and time whereas FinTech normally lacks intermediaries.

b. Despite of what can be commonly assured, some authors consider FinTech as being a separate industry from the traditional finance sector, as FinTech combines technology with finance and other industries.

c. Traditional finance and specially microfinance institutions offer as part of their value proposition a close emotion to their clients, which could be more difficult to FinTech to provide it, as FinTech is based on digital platforms.

d. An advantage for traditional finance is that they are regulated by financial associations and governments, which provide a feeling of security to their clients, while FinTech is currently not that stringently regulated.

4.4 Blockchain

Blockchain is one such powerful technology which has started a new era of internet – Internet-2.0 or Internet of Value. Blockchain is a decentralized database that consists of blocks that retain information, allowing the nodes composing the system to be aware of any data movement. This system is characterized by the fact that every part owns a copy of the ledger and cryptographic hashes to chain each block.

Some authors define it as a database, while others consider it as a system or network. It is composed of blocks that retain information or data, that are connected by ledgers and nodes in the form of a chain, the reason this system receives this name.

4.4.1 History of Blockchain

Generally, the beginning of blockchain is associated with the advent of Bitcoin in 2009. In a seminal paper Nakamoto (2008) mentioned this technology to explain a new decentralized peer-to-peer electronic cash system, which works on BLOCKS, which is CHANED with each other to transfer the movement of information on a real-time basis. Even Nakamoto did not use the word blockchain to express the term "time stamping", which is the core of this technology. At first, we found the mention of the term "time stamping" in 1991 (Haber & Stornetta, 1991), when it was used to mark the time on

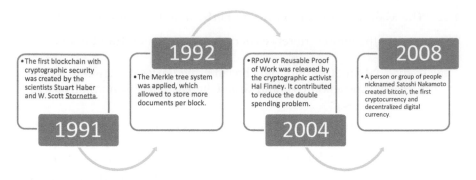

FIGURE 4.3
Most important stages in the history of blockchain.

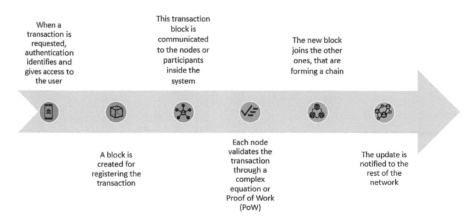

FIGURE 4.4
Process for blockchain functioning.

a digital document. As depicted in Figure 4.3, with time, this technology has developed, which we now know as blockchain, and came the year 2009, the year when Bitcoin was launched, which made this technology so popular that blockchain is considered as synonym of Bitcoin.

4.4.2 How Does Blockchain Work

Blockchain is a type of DLT or Distributed Ledger Technology, wherein each block in the network can save a record, such as the saving of the story of financial transactions (Meunier, 2018). Each block represents a record, and each one is stamped and encrypted for the correct data registration. Ensuring privacy and restricted access, blockchain is a permissioned network only for the known identities. Each user is given a private key for accessing their own block, and only the user that owns a block can modify its content, which will become immutable for the rest of users in the network (see Figure 4.4).

One of the main characteristics of blockchain is that it is decentralized, which means it lacks an authority inside the system, any ledger that is part of the network can visualize and access the data stocked in the blocks, and all the nodes are notified of data movements, so every part owns a copy of the ledger and cryptographic hashes to chain each block (Pilkington, 2016). As this is a real-time ledger, in case that any change in blocks is done, it is automatically updated for the rest to see it and validate the same.

4.4.3 Features of Blockchain

Blockchain is a peer-to-peer network; it doesn't require intermediaries. It is considered as an open system, where not only one person is in charge of the system, but it could be said that all the users of the network have a shared control on it.

Generally, due to the democratized feature of blockchain, this could be more open to people to participate, along with the feature of security and identity, which can be very helpful for preventing fraud that may occur to financial entities. Also, one of the most important applications is automating payments, which makes the process easier for the users. This can help the users to make sure that the payment was received, to have registers, and to be able to access documents with financial information from the other parts (Figure 4.5).

An additional element of blockchain is smart contracts; they can function when previous concerted conditions are presented, as they are characterized as being automated. The stipulations and conditions of the contract are completed in real time inside the blockchain, and when these stipulations and conditions are accomplished, they are fulfilled automatically (Golosova & Romanovs, 2018). This makes the system secure and trustworthy for the stakeholders. Figure 4.6 lists the advantages of blockchain technology.

Blockchain has been commonly associated with Bitcoin; however, nowadays with enterprises entering the digital world, it is not limited to just cryptocurrency. From sharing data with all the members of the company to tracking the route of a product to be delivered in supply chain management, blockchain has a variety of applications in business (Andolfatto, 2018).

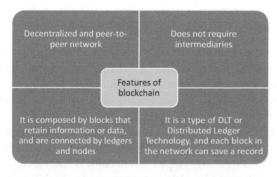

FIGURE 4.5
Main features and characteristics of blockchain. (Varma (2019).)

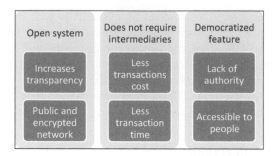

FIGURE 4.6
Advantages that blockchain offer to FinTech. (Fosso Wamba et al. (2020).)

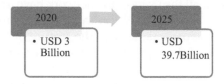

FIGURE 4.7
Global predicted growth of blockchain between the years 2020 and 2025.

Yimishiji is an online market for farmers in Shanghai, with an environmental focus on food.

Origin Trail, a European IT company, designed a blockchain system that controls food safety, maintains data integrity, and optimizes the connection with suppliers, which allowed them to be more efficient. The features of blockchain have made it an obvious choice for many industries, and it is used not only to provide various services but an underlying technology to invent other technologies. Figure 4.7 shows the exponential growth of Bitcoin over a period of time.

4.5 Applications of Blockchain in FinTech

Blockchain is an emerging technology, which has a great potential in financial services segment. It opens an opportunity for enterprises to have more developed technology in systems and databases for finance. According to PWC, 70% of businesses in the financial sector were willing to implement blockchain solutions. The expectations from blockchain were very high, and this could be like the ones that were when e-commerce or the internet appeared (Pilkington, 2016). Even though it is very difficult to have an exact prediction of what will really happen with blockchain in a couple of years, it is considered that it could be able to transform FinTech business models and the way they are currently managing their systems.

4.5.1 Cryptocurrency

Cryptocurrency was the first significant application in the history of blockchain, and the first cryptocurrency launched into the market was Bitcoin. After Bitcoin, other cryptocurrencies such as Ripple (XRP) were created, which uses blockchain as the underlying technology.

For exchanging cryptocurrency, crypto wallets are required for receiving and sending digital money, as well as to deposit money in the wallet. In terms of functionality, it is very similar to a bank account. Additionally, crypto wallets include a public key that contains the address of the wallet, so it will be visible to the other users, and it can be used to send coins.

The private key is like a password, and it serves for letting the user to have access to their crypto wallet. It could be said that its functionality is not too complex and is like money transfers but in a digital format (Stoica & Sitea, 2021).

Although some people do not have a lot of trust in cryptocurrency exchange as it lacks an official regulation and there are diverse opinions about this topic, platforms for digital currency exchange have become popular since the past years.

A positive aspect is that cryptocurrency can offer a unique way of payment in international transactions as it replaces the exchange of different currencies from each country. Having a variety of currency can be complex because of all further required conversions, so using only one type of currency for transaction could be simpler, without the need to recur to exchange rates. This means cryptocurrency is universal, so users from any part of the world can exchange money without being worried in changing their original country currency or imposed currency laws as per each country, along with the fact that until now, there is not an international law for cryptocurrency regulation or specific laws for transfer of fund (Chuen & Deng, 2017). Given this uncertainty or absence of clear-cut guidelines, it's evident, regulating entities and laws will be created and implemented at international level. In 2019, the European Commission on Regulatory Barriers to Financial Innovation suggested to consider the recent implementations of blockchain in FinTech, as part of the FinTech regulatory framework, for also offering new financial products soon.

As mentioned earlier, blockchain is majorly related with cryptocurrency. But blockchain is not limited only to cryptocurrency. New applications of this system are being implemented; FinTech is now offering new financial services that require more and more use of technology. Another reason is that the traditional financial sector does not want to be left behind, so they are considering the possibility to implement blockchain to digitalize their services and processes.

4.5.2 Banking

One of the segments that has been significantly impacted by blockchain in finance, wherein new innovative services can be developed digitally, which expands the market of FinTech, is banking (Shah & Jani, 2018). In fact, it is banking industry which uses 30% of blockchain application as blockchain features support in fulfilling the requirements of banking sector. Quicker KYC for clients, reduced settlement period, digitization of documents, etc. are few of the requirements of banking business, which is very well taken care by blockchain technology. Keeping an accurate record of transactions is essential for banking and accounting firms, and maintaining the trail of transactions is the prime feature of blockchain. It simplifies and fastens the payment process as there is no requirement of intermediary under this system (Broome, 2019). As all the transactions are as per the already accepted parameters and are verified by existing members, there is absolutely no chance of fraud. If anyone even tries to defraud the company, he/she will have to re-write the already existing blocks, which is almost impossible; also, this activity will alert everyone regarding the suspicious activity. Banks are increasingly using blockchain for achieving digitalization for their processes, as they require to have faster answers for their clients, and they need to offer their service for it to be available in multiple devices, including the mobile ones, and have an international reach (Ozaee & Sohrabi, 2017). Furthermore, the banking industry is constantly adapting to the industry and people's needs, and they must develop new products and services to maintain their position in the market with innovation and promoting digital transformation. An example of this is to focus on new niches, such as voice banking (Stoica & Sitea, 2021).

4.5.3 Data Registration and Identification

Data registration and identification is one of the very important features of financial services. With blockchain, businesses register and save data from employees, customers, and transactions of the business; the members can have access to it and make further analysis. This will help to establish a better relationship with them and use their data to make action plans according to the requirements of the organization (Buitenhek, 2016).

Blockchain offers the possibility of having a personal identification (ID) to access the data. As financial data is sensitive, blockchain is looking forward to assuring security for users and to avoid hackers to steal information.

4.5.4 Payments

Payments can be considered as one of the most common blockchain applications. Nowadays, payments are faster than before. Due to the technological advancements, payments are processed almost instantaneously, as because of absence of intermediary, processing time is reduced significantly. With blockchain, the nodes that connect the blocks with the financial information automatically accepts or deny the payment, making the payment process more efficient and easier (Peter & Moser, 2017). This allows the user to not only save time but also save money. Also, the system protects data integrity, as each node will have to be accessed by its owner for being modified. Furthermore, cryptocurrency payments have been increasing since the last few years, even though they are not officially regulated by governments.

4.5.5 Trade Finance

Trade finance is an example in which blockchain is being applied. As, in blockchain, presence of an intermediary is not required, the processes become more efficient and easier for customers. This leads to a lower cost because there won't be a cost applying for intermediaries. Also, it can simplify the trade finance process for all the involved parties, as it reduces the use of manual documentations processes and makes visible for everyone the details of the transactions and the route of the products (Bogucharskov et al., 2018).

For avoiding fraud and assuring the fulfilment of the agreement conditions from all the parties, smart contracts is appended on blockchain, which automatizes the corresponding payment if the conditions have been accomplished, which can also contribute to increase trust, as both parties now have a clear vision of the process and justice if the conditions have not been met.

4.5.6 Insurance

Insurance is one such segment where the blockchain technology is used extensively. This pandemic has shown the people at large the importance of insurance services. Blockchain helps here to identify the eligibility of the insured, sign the contract, and to ensure the amount reaches the true beneficiary (Chuen & Deng, 2017). As the segment of FinTech is exponentially growing, newer services and uses of technology are coming up to expand the market, and blockchain is one of the reasons of the growth. Because of patient confidentiality clause, insurance providers often don't have access to patients' full medical history. Lack of data can lead to insurance claim denials, which costs hospitals $262B yearly and is cited as a significant factor in rising healthcare costs. A single patient will typically see multiple doctors and specialists over the course of his life. Because there are so many different parties involved in healthcare, it's difficult to share and coordinate sensitive medical data between them. Blockchain technology can encrypt patient information, facilitating the transfer of information while still protecting patient privacy.

4.5.7 Financial Inclusion

This technology can contribute a lot to achieve financial inclusion with the aim of making financial products and services accessible to bottom of the pyramid. DLT allows financial institutions to create synergies through their ecosystem, know their customers properly, detect fraudulent transactions reliably, and diversify their risks more effectively.

4.5.8 Investments

Blockchain, appended with smart contracts, can be useful for investments; as when someone decides to invest, there are already clear conditions for both parts. Smart contracts will help to make the process simpler and help to optimize it, along with rules and stipulations. Blockchain can help these rules to be enforced with the network, for checking that all conditions have been met before the investment takes place (Chuen & Deng, 2017).

Smart contracts are characterized to be efficient and to keep privacy, so it will be very helpful. So, the process will be faster, and the financial data can be safe, thanks to the encryption protocols for security. It also can help to improve the relation with the investor, as both agreement parts will be clearer.

4.5.9 Credit and Alternative Lending

Blockchain is considered as a peer-to-peer network, and this system works very well with credits too. In contrast with traditional finance, criterion such as credit score and credit history won't be considered, which increases the possibility of more people to have access to a loan, as the selection criterion would be a global credit score, obtained by decentralized historical payments register (Bogucharskov et al., 2018).

This opens the possibility of alternative lending supported by blockchain, which is characterized for being faster, with less prerequisites when applying for a credit or a loan and having a lower cost. It can also be considered as being more secure, due to cryptography.

4.5.10 Tokenization

Tokenization of digital goods and assets is another application of blockchain. A token could be defined as the representation in digital format of a property or good in the commerce sector, and in blockchain it is considered as a means for payment as well as an investment tool or has multiple marketable uses. In other words, a token can represent a variety of assets. Smart contracts can be used in tokenization, in the cases of transfers. Other movements such as acquisition investment and commercialization can be made, as well as to tokenize assets. An added value is that tokens can consist of transparency, automated payments, and exposure to potential investors. Also, there exist secondary markets for security tokens, and in general, tokenization can help physical assets to be liquid in the market. Another type of tokenization is property tokenization, which involves transferring a certain property. The different parts that form the property represent a certain number of tokens and the whole is a property. Property tokens are asset or security tokens. For example, collaborative economy may be applied, and a variety of investors can be reached to let them know about it.

4.6 Successful Cases of Blockchain in FinTech

There are many successful cases of application of blockchain technology in FinTech. Few of them are listed below.

4.6.1 HSBC

The HSBC bank started implementing blockchain technology in 2018, and one of its main uses in global trade is providing letters of credit. Letters of credit are a guarantee of a financial institution that seller will get his due amount. Globally trade comes with their own set of complexities owing to exchange rate, different banks, etc. which make the payment process cumbersome and time consuming. HSBC implemented this technology to track the transaction and also has begun to use the contour platform with other seven banks, for reducing the time spent in trades along with providing capital efficiency to their customers.

Regarding foreign exchange transactions, the bank uses a platform "HSBC FX Everywhere" that provides more efficiency to the payments due to DLT, and they have made approximately more than 2 million trades. Additionally, for processing the payments of various participants, the bank implemented HSBC's On-Chain Payment, that consists in relating the different ecosystems with the blockchain network, along with smart contracts. Recently HSBC has completed its first Bond issuance worth S$5.5 million and a follow-up issuance of S$100 million in a streamlined manner, all thanks to DLT.

For private placements, Digital Vault blockchain was created, as a custody-based platform, for allowing clients from all over the world to have access to their private assets in the digital platform of HSBCnet. Normally this type of transaction is not in a digital format as there is no standard protocol for it, although HSBC wanted to start digitalizing their processes and customers also wanted to have a real-time access to their private placements. Along with HSBC FX Everywhere, these platforms have received awards in the category of transaction banking in 2020 by Forbes.

Furthermore, thanks to blockchain implementation, HSBC was considered in the Forbes Blockchain 50 list, as one of the enterprises that was obtaining benefits from their use of blockchain, such as accelerating processes and improving revenues obtained. For the bank, blockchain contributed to make the processes to be more efficient, because the presence of an intermediary was not necessary, which can lead to a lower cost. It can also help to strengthen data transparency and to improve security in the different networks that own the bank. Here, their groups of interest that make business transactions were involved, and they will start promoting more technology acquisition.

4.6.2 U.A.E. Blockchain Strategies

This case is not about a company, but of a country that applied blockchain as part of a national strategy for FinTech development. United Arab Emirates is a country located in Western Asia and known for its development in technology. According to the U.S.-U.A.E. Business Council, U.A.E. can be considered as an important FinTech hub in Middle East and North Africa or MENA region. This is because approximately, there were 46% of FinTech start-ups in the MENA region in 2019, supported with the help of the Abu Dhabi Global Market and the Dubai International Financial Centre and their accelerator programmes, funding, and consulting services, as well as the focus on an economy for entrepreneurship.

In direct relation to blockchain, in the year 2018, the United Arab Emirates government created the Emirates Blockchain Strategy 2021. Its main objective is to register and capitalize 50% of the government transactions to the blockchain platform until 2021 as the goal deadline, for it to also conduce further transactions. This mainly considers FinTech start-ups because this strategy is willing to include e-payment services. Its focus is based on four pillars, which are "citizen and resident happiness, government efficiency, advanced legislation, and global entrepreneurship".

One of its main advantages is the application of innovative technology as part of government transactions, which contributes a lot in the technological development of the country, and the government passes the benefit to the private sector directly. Cybersecurity of digital transactions continue improving because the information storage on blockchain cannot be manipulated. Approximately 1 billion dirhams (U.A.E.'s local currency) was saved, and requirement of printed documents and the unlimited work hours for this activity were removed.

Another strategy is the Dubai Blockchain Strategy, which has the objective of transforming Dubai, one of the most popular cities in U.A.E., to be being recognized as the first blockchain powered government in the world and an important national economic driver. For achieving this, they focus on three pillars: government efficiency, industry creation, and international leadership, following the vision of Hamdan bin Mohammed al Maktoum, the Crown prince of Dubai.

Furthermore, the Global Blockchain Council in Dubai has been created by the Dubai Future Foundation, conformed by businesses in financial and non-financial sector and government entities, with the aim of improving the use of blockchain.

4.6.3 Ripple

Ripple is an example of cryptocurrency and payment system that was created in San Francisco, California in 2012. Today, it is considered as one of the best payment systems. It consists in a payment and transfer platform for individuals and SMEs (Small and Medium-Size Enterprises) that do not require intermediaries, so users such as banks and enterprises can transfer money directly, along with cryptocurrency exchange, from any part of the world.

Additionally, it offers personalized integrations, payment track, and efficiency, focusing more on reliability and lower costs. As it is a decentralized platform, their clients are able to make payments worldwide and not paying a prefunding cost, thanks to the On-Demand liquidity service and the use of the XRP ledger and leverage cutting-edge blockchain. Also, the company suggests that the decentralized nature of their platform will make their service more accessible for people and helping achieve financial inclusion.

Another offered service is the line credit, in which SMEs can complete a payment with the credit line and finish to return it later.

4.7 Issues with Blockchain

Even though there are various advantages and benefits that blockchain can provide to FinTech, there are still some problems and issues that may be involved with its implementation.

"Blockchain's sluggish transaction speed is a major concern for enterprises that depend on high-performance legacy transaction processing systems", to help understand this, we can compare Bitcoin which processes merely 4.6 transactions per second as compared to Visa which does 1,700 transactions on an average; this gives rise to the problem that blockchain technology will not work for users or organizers who demand a fast performance. Also, since blockchain technology is highly standardized, it won't be an ideal choice for a customized deal between two or more parties. By and by, the storage cost is staggeringly high.

A characteristic of blockchain is that it can possess and have access to personal information, which can reduce privacy and freedom of information. Additionally, if someone loses its key, it can be difficult to recover the access to the account and consequently the saved information. This could be a disadvantage if a user of enterprise has all their information in this system, or valuable information, in which the loss of this data can bring serious consequences to the user or business.

Because there exist a lot of frameworks about the functionality of blockchain, and as it is a decentralized system, it involves a lack of standardization. This can lead to a disorganization inside the network, along with an incorrect interpretation of the actions taken inside it. It also lacks an official regulation, which can cause a feeling of insecurity and informality in some users, especially for ICO or Initial Coin Offerings. In this topic, there exist less regulations than in cryptocurrencies, which can mean to be a higher risk for investors.

Nowadays, blockchain is achieving a lot of popularity. Because this system can be considered as being in trend currently, a lot of enterprises from different sectors are looking to implement this system. This causes blockchain growth to be fast, which leads to the fact that every company wants to develop their own blockchain network in an individual way. The origin of this is that there are not well-developed standards for blockchain networks, which consequently can cause the lack of interoperability between systems.

Also, this could derive in people not knowing too much about how to use the system or being more exposed to suffer a loss of their data. An example is the phishing technique, which consists in the user to be misled for then sharing a password and stealing private information.

The high expectations from blockchain can cause the network to be in a bubble situation. This means that similarly to what happened with the internet when it was launched. The stock price of the enterprise, that implement blockchain technology in their processes, increases. These enterprises must probably have a high investment for implementing this system, if the bubble explodes, then the prices can decrease, and the enterprises may be less sustainable in financial terms.

In the sustainable aspect, it could be said that the development of blockchain networks is not friendly with the environment, as it not only has a high economic cost but also a high energy cost. This occurs when the blockchain system uses the proof-of-work or PoW consensus method for preventing attacks. In it, a difficult mathematical problem must be resolved, and this involves using machines that have a high energy consumption and posterior cost. Furthermore, blockchain can have an operational risk, which reduces in possible deficiencies in system, as this system is relatively new, and its security aspect can be improved. For example, this can lead to a probability of the system to get hacked, which can lead the deficiencies in the system to get the hackers to access information by data manipulation. Also, when a financial company outsources their services, this may occur, as they do not have a direct control of it, or when the users do not have a correct use of the platform, which can have a negative effect on the business.

Additionally, the fact that blockchain has not many years in the industry can translate into a strategic risk for businesses. This occurs because the enterprise must evaluate in deepness if this system is the most appropriate for them, according to the business nature, objectives and the product or service they offer can be adequate to be delivered by this network. Also, because blockchain is a P2P or peer-to-peer system, this means that various entities will probably participate in the network, so it must be evaluated if this aspect will have a significant impact on the business.

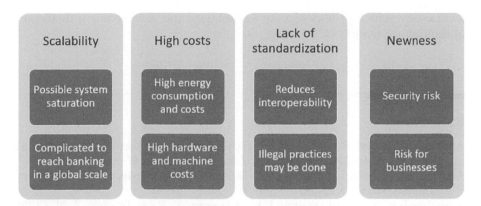

FIGURE 4.8
Problems and issues that can arise due to blockchain use. (Fosso Wamba et al. (2020).)

One of the main issues of blockchain is scalability, which means the capacity of storing a large quantity of data. Blockchain currently does not have a big scalability, so it will be very difficult for it to reach banking in a global scale as it will demand more improvements on the existing network. For example, if there is a congestion or too much data movements occurring in parallel, the system will be slower and less efficient, as bigger data volumes as being handled at the same time (Joshi et al., 2018).

Data privacy is important for protecting the users' personal information, so even if blockchain is offering that the profiles will maintain anonymous, it is still necessary for the network to implement this option (Yeoh, 2017). Since blockchain can be accessed worldwide, each country may have an independent opinion and laws regarding data privacy, so implementing a global data privacy regulation is difficult (see Figure 4.8).

4.8 Conclusion

Blockchain is a peer-to-peer network, and even though it was launched as the system that will support Bitcoin exchange, nowadays many businesses from various sectors are starting to get interested in implementing blockchain technology as part of their processes. The benefits they would like to obtain are mainly to digitalize their processes and be more efficient, as well as the fact that blockchain can have more uses and applications, in addition to supporting cryptocurrency exchange. Most authors in this theme are forecasting that blockchain will have a growth in the following years, as it has also increased since the last years, although it is still uncertain if this growth will concretely achieve, or it will remain as a high expectancy that finally won't be carried out. But, in current situation, blockchain is considered a technology, which provides many advantages, and is seen as a transformer of the finance world.

References

Andolfatto, D. (2018). Blockchain: What it is, what it does, and why you probably don't need one. *Federal Reserve Bank of St. Louis Review,* 100(2), 87–95.

Arner, D. W., Barberis, J., & Buckley, R. P. (2016). 150 years of FinTech: An evolutionary analysis. *JASSA,* 3, 22–29.

Bogucharskov, A. V., Pokamestov, I. E., Adamova, K. R., & Tropina, Z. N. (2018). Adoption of blockchain technology in trade finance process. *Journal of Reviews on Global Economics,* 7, 510–515.

Broome, L. L. (2019). Banking on blockchain. *NCJL & Technology,* 21, 169.

Buitenhek, M. (2016). Understanding and applying blockchain technology in banking: Evolution or revolution? *Journal of Digital Banking,* 1(2), 111–119.

Chang, M. C., & Park, D. (2020). How can blockchain help people in the event of pandemics such as the COVID-19? *Journal of Medical Systems,* 44(5), 1–2.

Chen, Y., & Bellavitis, C. (2019). Decentralized finance: Blockchain technology and the quest for an open financial system. Stevens Institute of Technology School of Business Research Paper.

Chuen, D. L. K., & Deng, R. H. (2017). *Handbook of Blockchain, Digital Finance, and Inclusion: Cryptocurrency, FinTech, Insurtech, Regulation, Chinatech, Mobile Security, and Distributed Ledger.* Academic Press: Cambridge, MA.

Deshpande, D. S., Nirala, A. K., & Salau, A. O. (2021). Implications of quantum superposition in cryptography: A true random number generation algorithm. *Smart Innovation, Systems and Technologies,* vol. 196, pp. 419–431. Springer, Singapore. DOI: 10.1007/978-981-15-7062-9_41

EY (2019). Global FinTech Adoption Index 2019. https://www.ey.com/en_gl/ey-global-fintech-adoption-index.

Fosso Wamba, S., Kala Kamdjoug, J. R., Epie Bawack, R., & Keogh, J. G. (2020). Bitcoin, Blockchain and FinTech: A systematic review and case studies in the supply chain. *Production Planning and Control,* 31(2–3), 115–142. doi: 10.1080/09537287.2019.1631460.

Gai, K., Qiu, M., & Sun, X. (2018). A survey on FinTech. *Journal of Network and Computer Applications,* 103, 262-273.

Goldstein, I., Jiang, W., & Karolyi, G. A. (2019). To FinTech and beyond. *The Review of Financial Studies,* 32(5), 1647-1661.

Golosova, J., & Romanovs, A. (2018). The advantages and disadvantages of the blockchain technology. *In 2018 IEEE 6th Workshop on Advances in Information, Electronic and Electrical Engineering (AIEEE),* pp. 1–6. IEEE.

Haber, S., & Stornetta, W. (1991). How to time-stamp a digital document. *Advances in Cryptology.* Lecture Notes in Computer Science, vol. 537. Springer, Berlin, Heidelberg. doi: 10.1007/3-540-38424-3_32.

Hashemi Joo, M., Nishikawa, Y., & Dandapani, K. (2019). Cryptocurrency, a successful application of blockchain technology. *Managerial Finance,* 46(6), 715–733.

Joshi, A. P., Han, M., & Wang, Y. (2018). A survey on security and privacy issues of blockchain technology. *Mathematicalfoundationsofcomputing*, 1(2), 121–147. doi: 10.3934/mfc.2018007.

Khurshid, A. (2020). Applying blockchain technology to address the crisis of trust during the COVID-19 pandemic. *JMIR Medical Informatics*, 8(9), e20477.

Meunier, S. (2018). Blockchain 101: What is blockchain and how does this revolutionary technology work? In: Marke, A. (ed.) *Transforming Climate Finance and Green Investment with Blockchains*. Academic Press: Cambridge, MA, pp. 23–34.

Nakamoto, S. (2008). Bitcoin: A peer-to-peer electronic cash system. *Decentralized Business Review*, 21260.

Ozaee, A., & Sohrabi, S. (2017). The role of financial technology and their effect on banking. *Revista QUID (Special Issue)*, 1819–1826.

Peter, H., & Moser, A. (2017). Blockchain-applications in banking & payment transactions: Results of a survey. *European Financial Systems*, 141, 141.

Philippon, T. (2016). The fintech opportunity (No. w22476). National Bureau of Economic Research.

Pilkington, M. (2016). Blockchain technology: Principles and applications. In: Xavier Olleros, F., & Zhegu, M. (eds.) *Research Handbook on Digital Transformations*. Edward Elgar Publishing: Cheltenham, pp. 225–253. doi: 10.4337/9781784717766.00019.

Roth, A. E. (2015). *Who Gets What—And Why: The New Economics of Matchmaking and Market Design*. Houghton Mifflin Harcourt: Boston, MA.

Shah, T., & Jani, S. (2018). *Applications of Blockchain Technology in Banking & Finance*. Parul University, Vadodara, India. doi: 10.1007/s41471-020-00090-5.

Stoica, E. A., & Sitea, D. M. (2021). Blockchain disrupting FinTech and the banking system. *Proceedings*, 74(1), 24. doi: 10.3390/proceedings2021074024.

Treleaven, P., Brown, R. G., & Yang, D. (2017). Blockchain technology in finance. *Computer*, 50(9), 14–17.

Varma, J. R. (2019). Blockchain in finance. *Vikalpa*, 44(1), 1–11. doi: 10.1177/0256090919839897.

Yeoh, P. (2017). Regulatory issues in blockchain technology. *Journal of Financial Regulation and Compliance*. 25(2), 196–208. doi: 10.1108/JFRC-08-2016-0068.

Zavolokina, L., Dolata, M., & Schwabe, G. (2016). The FinTech phenomenon: Antecedents of financial innovation perceived by the popular press. *Financial Innovation*, 2(1), 1–16. doi: 10.1186/s40854-016-0036-7.

5

Blockchain for E-Governance and Tracing Fake News on Social Media

Susheela Dahiya and Keshav Kaushik

University of Petroleum & Energy Studies (UPES)

CONTENTS

5.1 Blockchain for E-Governance

Electronic governance or e-governance is the information and communication technology (ICT)-enabled mode of delivering government services, exchange of information, communication transactions, and integration of various stand-alone systems and services. In the last decade, government services have evolved from paper-based to digital services. The digital services or e-services (electronic services) require only a limited human interaction and helps in providing government services in a convenient, efficient, and transparent manner along with a fast response. There are four types of interactions that can take place in e-governance: Government to Citizen (G2C), Government to Business (G2B), Government to Government (G2G), and Government to Employees (G2E).

With e-governance, government can easily upgrade the efficiency and quality of the services. E-governance can also reduce the paperwork and improve the delivery of government services. Also, citizens can easily access the legitimate information about government services as per their convenience which can help in increasing transparency and reducing corruption. Because of all these advantages, every passing year more and more countries are embracing e-governance technologies. As per the E-Government Survey conducted by Department of Economic and Social Affairs, United Nations, in 2020, 193 countries are actively providing e-governance applications [1]. Out of these 193 countries, 63 countries including India are providing almost all features and 22 countries are providing limited features considered for

DOI: 10.1201/9781003193425-5

the assessment in the survey. Also, 162 out of 193 countries are offering at least one online transactional service like paying utility bills, applying for marriage certificate, birth certificate, driving license, passport, visa, etc. The online transactional service deals with private and confidential data of users which require to address data security and privacy issues.

E-governance deals with a large amount of confidential data of citizens as well as of government. Blockchain technology is very useful in the areas in which processing of a large amount of personal and protected data is involved [2]. With the help of blockchain technology, e-governance can provide more secure and transparent public services or exchange of information. In the last few years, record keeping, smart contracts, and value transfer are the three main characteristic areas of blockchain technology that has been adopted in e-governance system. There are seven e-government services which are supported by blockchain [3]:

i. **Authentication**: Blockchain based authentication is very useful to verify the digitally identity of a person. Blockchain based authentication is required to protect the passwords and other personal/private data of a user. The types of blockchain that can be used for the authentication are consortium, Hyperledger private and public permissioned blockchain.

ii. **e-Voting**: Modern culture is increasingly embracing the practise of online voting. It has a significant chance of lowering administrative expenses and raising participation rates. Voters can cast their ballots from any location that has an Internet connection, eliminating the need for printed ballots or accessible polling places. Online voting alternatives are nevertheless considered with considerable care because they pose new risks despite these advantages. Voting tampering on a massive scale may be possible due to a single vulnerability. When utilised in elections, electronic voting systems must be reliable, precise, secure, and practical. However, possible issues with computerised voting methods can restrict implementation. Blockchain technology was developed to address these problems, and it now provides decentralised nodes for electronic voting. Electronic voting systems are created using blockchain technology primarily because to the benefits of end-to-end validation. With decentralized, non-repudiation, and proper security features, this innovation is an excellent substitute for conventional electronic voting systems.

iii. **Data Sharing**: Data protection is very important while sharing it with another user (a person or department). Blockchain can be used to authenticate each user and prevent counterfeit of data.

iv. **Land Property Services:** The biggest asset class worldwide is real estate. Businesses and real estate experts are starting to realise how revolutionary blockchain technology may be for streamlining payments, increasing access to real estate capital, and optimising sales of commercial and retail properties. Immutable documents and a transaction history that can be audited can be provided by blockchain technology to maintain property or land titles.

v. **E-Delivery Services:** Although the Indian e-commerce market has faced many difficulties, it has also been searching for reliable technology assistance to handle these difficulties. Thankfully, blockchain technology for e-commerce is prepared to propel organizations forward. The finest use of blockchain technology in e-commerce is smart contracts, which are used to control the provision of logistical and payment processing operations. The technology also simplifies the entire process and makes it easier to integrate with other e-commerce and logistics management systems. Blockchain in e-commerce boosts operational efficiency by recording data, and it also assures supply chain visibility. The firms can fully visualize the product's processing and origin thanks to technology.

vi. **Human Resource Management:** Blockchain can automate and secure payments to workers, independent contractors, and vendors, streamlining the payroll process. Cross-border transactions, which contractors and "gig economy" employees frequently need, are an early use that was initially presented a few years ago by various businesses. They occasionally don't have bank accounts, which are often necessary for automated payroll system payments. Blockchain payroll methods circumvent local rules and IT security measures that can thwart conventional electronic payments. Major HR software providers are participating.

vii. **Government Contracting:** Blockchain based systems can also be used for real time online monitoring of the different government sectors where government advertise tenders and award the contract after evaluation of the tender bids. The government can prepare and uploads the tender documents online and the interested parties can apply as soon as the contracts are open by government. Incorporation of blockchain in e-Tendering automate the entire process and makes it more transparent. At every step of the e-Tendering process i.e. Preparing and publishing of the tender documents, submission of the bid by the bidders, opening of the tender, bid evaluation and award of the contract; a new block of the data is created and added to the existing blockchain which is permanent and unalterable [4]. Therefore, no alteration can be done in the submitted bids. The participants/bidders are not required to physically visit the government offices. It also helps in streamlining and completing the whole process in a timely manner.

5.1.1 Security and Privacy Concerns in e-Governance

Because of the involvement of personal data, there are many security and privacy concerns that should be considered while. Blockchain technology is used by governments and public organizations to do rid of centralized, siloed systems. Blockchain networks provide more secure, adaptable, and economical architectures than current systems, which are by nature unsecure and expensive. A blockchain-based digitalization may increase confidence and accountability while securing data, streamlining procedures, and reducing fraud, inefficiency, and abuse. Individuals, companies, and governments all share resources on a distributed ledger that is encrypted using cryptography in a blockchain-based model of government.

5.2 Blockchain for Tracing Viral or Fake News on Social Media

Fake media, sometimes known as the Web of Dishonest Journalism, has appeared in a number of spheres of information society, particularly politics, media, and social media. Because the media's reputation is often undermined, drastic steps are essential to avoid future erosion. Considering today's Artificial Intelligence and Deep Learning breakthroughs, Internet of Fleet Management Things (IoFMT) is becoming increasingly widespread; yet, such learning concessions may be severely constrained. It is vital to offer proof of authenticity in order to identify provenance and integrity of any digital output. A blockchain is a distributed ledger technology. In order to help with the issue, a potential new decentralized safety infrastructure has been presented. The technological component of false media is critical in a data-driven world. Nevertheless, various blockchain-based authentication systems have been proposed. The bulk of available research, on the other hand, is based on erroneous post-incident views.

When working with a large amount of data and services, it might be difficult to maintain safety and confidentiality at the same time. Any item of content that is not true in any way, shape, or form is a false piece of information. Claims that are partly or wholly untrue in some way are the polar opposite of genuine assertions. Information that is posted to social media or sponsored via advertising is also a good way to get it out there. In combination to vagueness, material accuracy, opinion impact, poor judgment, and voting patterns, misleading media information may inflict bodily injury.

Blockchain technology has paved the way for the development of decentralized applications in which security is paramount. Any transaction that has ever taken place is forever recorded here. Over the years, several shady sources have published phony and appealing news items. Such information cannot be verified since there are no governmental processes in place. As a result, these untrustworthy sources may publish anything they choose, causing social upheaval in certain circumstances. Because of the widespread availability of the internet and social media in recent years, false information has traveled faster than ever before. Fake news might be more appealing than actual news in certain instances. As a result, individuals get misled. In Ref. [5], the authors highlighted about how to identify false news on social media using the benefits of blockchain's peer-to-peer network ideas.

In Ref. [6], the authors provided a high-level description of a blockchain-based architecture for preventing false news, as well as the different design difficulties and factors to consider while developing such a system.

While social networking sites make modern-day communication much easier, they have also introduced new problems to real-world application, such as the viral dissemination of fake information with harmful intent. A naïve blockchain [7] and watermarking-based social media system is presented in this study to restrict the spread of bogus news. To address the current issues in this industry, we propose a new blockchain paradigm. Furthermore, by locating the source or origin of bogus news on social media, the unique method may aid in preventing the propagation of fake news.

Blockchain [8] has emerged as a major technology for maintaining the network's event logs, since it is built on a distributed ledger. For IoT devices, blockchain enables a secure, decentralized, and permissionless environment. Fake news [9] and rumors concerning COVID-19 are widely disseminated on social media platforms. Blockchain has the potential to significantly reduce the spread of bogus news. Several projects have already been undertaken in this approach, with the goal of implementing a blockchain-based social media network to identify the source of bogus news. Content analysis, artificial intelligence, and machine learning are other key study fields for analyzing false news since they can identify fake news based on published material.

The research in Ref. [10] intends to give an understanding of how blockchain technology may be used on social media to increase public confidence in legitimate news transmitted via popular social media sites, as well as to assess the source's veracity and prevent the spread of false news. This study combines blockchain technology with superior artificial intelligence in social networking sites to check the legitimacy of news material. Fake news has grown to be a big issue online, with far-reaching implications. Open access and uncontrolled social media platforms are common attack vectors for disseminating false information. Numerous naïve methods have been offered to combat this issue, including a blockchain version of a news feed to identify reality from falsehoods. The magnitude of social networking sites and the architectural structures of blockchain, on the other hand, provide various additional problems that obstruct the application of such solutions in the actual world. In Ref. [11], the authors proposed a new blockchain architecture that addresses the current issues while also limiting the propagation of bogus news throughout the network.

Users may now publish news items to social networking sites and unregulated sites without demonstrating their truth, while conventional news outlets followed rigid norms of behavior to validate reports. Because there are no determinants of the authenticity of such news pieces on the internet, a unique way to determining the soulfulness quotient of unconfirmed news items using technology is required. The research in Ref. [12] proposes a dynamic model with a secure voting system in which news reviewers may give input on news items, and a probabilistic mathematical formalism is utilized to forecast the veracity of the news item depending on the responses received. ProBlock, a blockchain-based approach, is developed to assure the accuracy of information disseminated.

5.3 Conclusion

In this chapter, the authors have presented the applications of blockchain in the domain of e-governance and tracing fake news on social media. As we all know that people are using social media for posting their sentiments, but people are also using social media to spread the fake news. Therefore, the blockchain technology is used in tracing the source of the fake news on social media, thus helping the police officials and cybercrime investigators. This chapter is helpful for the blockchain enthusiasts, researchers, and the students who are interested in learning the various applications of blockchain technology in e-governance and social media.

References

1. Department of Economic and Social Affairs, United Nations (2020). UN E-Government Survey 2020. https://publicadministration.un.org/egovkb/en-us/Reports/UN-E-Government-Survey-2020.
2. D. Markusheuski, N. Rabava, and V. Kukharchyk (2017). Blockchain technology for e-governance. *In The Conference of Innovation Governance in the Public Sector.* Kazan, Republic of Tatarstan, Russia.
3. Lykidis, I., Drosatos, G., & Rantos, K. (2021). The use of blockchain technology in e-government services. *Computers, 10*(12), 168.
4. O. Pal & S. Singh (2019). Blockchain technology and it's applications in e-governance services. *International Journal of Recent Technology and Engineering (IJRTE),*8(4), pp. 5795 - 5802. ISSN: 2277-3878.
5. S. Paul, J. I. Joy, S. Sarker, A. A. H. Shakib, S. Ahmed, & A. K. Das (2019). Fake news detection in social media using blockchain. *The 7th International Conference on Smart Computing and Communications (ICSCC 2019).* doi: 10.1109/ICSCC.2019.8843597.
6. A. Qayyum, J. Qadir, M. U. Janjua, and F. Sher (2019). Using blockchain to rein in the new post-truth world and check the spread of fake news. *IT Professional,* 21(4), 16–24. doi: 10.1109/MITP.2019.2910503.
7. A. D. Dwivedi, R. Singh, S. Dhall, G. Srivastava, and S. K. Pal (2020). Tracing the source of fake news using a scalable blockchain distributed network. *Proceedings of 2020 IEEE 17th International Conference on Mobile Ad Hoc Smart System (MASS 2020),* pp. 38–43. doi: 10.1109/MASS50613.2020.00015.
8. A. Dhar Dwivedi, R. Singh, K. Kaushik, R. Rao Mukkamala, and W. S. Alnumay (2021). Blockchain and artificial intelligence for 5G-enabled Internet of Things: Challenges, opportunities, and solutions. *Transactions on Emerging Telecommunications Technologies,* e4329. doi: 10.1002/ETT.4329.
9. W. B. Demilie, and A. O. Salau (2022). Detection of fake news and hate speech for Ethiopian languages: a systematic review of the approaches. *Journal of Big Data,* 9, 1–17. DOI: 10.1186/s40537-022-00619-x.

10. K. Kaushik, S. Dahiya, R. Singh, and A. D. Dwivedi (2020). Role of blockchain in forestalling pandemics. *In Proceedings of 2020 IEEE 17th International Conference on Mobile Ad Hoc and Smart Systems (MASS 2020)*, pp. 32–37. doi: 10.1109/MASS50613.2020.00014.

11. M. Saad, A. Ahmad, and A. Mohaisen (2019). Fighting fake news propagation with blockchains. *2019 IEEE Conference on Communications and Network Security (CNS 2019)*. doi: 10.1109/CNS.2019.8802670.

12. E. Sengupta, R. Nagpal, D. Mehrotra, and G. Srivastava (2021). ProBlock: A novel approach for fake news detection. *Cluster Computing*, 24(4), 3779–3795. doi: 10.1007/S10586-021-03361-W/TABLES/6.

6

Integration of Blockchain Technology and Intelligent System for Potential Technologies

Manish Thakral, Rishi Raj Singh, and Keshav Kaushik

University of Petroleum & Energy Studies (UPES)

CONTENTS

6.1 Introduction

In recent years, digital currencies like bitcoin, blockchain, and artificial intelligence (AI) have become increasingly well known due to their crucial roles in technological innovation and industrial change. In 1959, the Harvard Society presented an idea on AI known as the E-chain, which was created by the Harvard Institute of Technology [1]. AI is a technology that aims to simulate, extend, and enhance human intelligence. A definition of AI is as follows: machine learning advances have opened up the way for developing AI at an accelerated pace, which is expected to continue in the future. AI is particularly helpful in these multiple sectors since it facilitates a more informed decision-making process, especially in situations of analysing, predicting, and assessing situations. Blockchain technology, which was introduced by Satoshi Nakamoto when bitcoin was created, has existed for a long time [2]. Consider the notion of blockchain as similar to the notion of a distributed ledger as we think of an E-chain. When several entities engage with each other without a trusted third-party present to facilitate agreement, a decentralised consensus process can be utilised. Using blockchain technology in an untrusted distributed system allows for generating and verifying transactions at a lower cost than conventional methods, which is beneficial. Academics and researchers increasingly pay attention to blockchain technology for precisely this reason [3] (Figure 6.1).

Both AI and Ethereum have their own set of advantages and downsides, which are discussed more below. It is conceivable that the blockchain may create difficulties in the energy sector. Blockchain technology offers many benefits, like continuous on-demand consumption, limitless scalability, top-notch security, and complete anonymity, among others. While AI has challenges in terms of interpretability and efficacy, these challenges are shared by other areas of AI as well. It is possible that both of these research paths are interconnected and mutually dependent on one another. In addition to being prepared to place demands on analytics, security, and trust, natural integration offers a variety of other benefits for all parties involved. For example, therefore, they will be more equipped to improve the lives of their peers while also empowering one another as a consequence. Algorithms, strategies, and networks, to mention a few of the most essential components of AI, are among the most important components of computer science. The blockchain will put to the test not only power and data, but also an island of data, and then it will make use of the current flow of compute, processing power, and data resources flexibility, as well as anonymisation, in order to achieve its goal of decentralisation. As an added benefit, blockchain ensures that data is completely secure, trustworthy, and auditable, while also addressing the

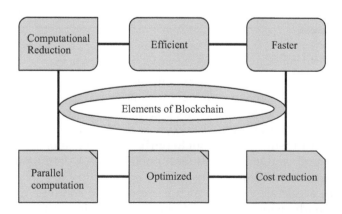

FIGURE 6.1
Depiction of blockchain elements with respect to AI.

láck of transparency and evidence requirements in AI applications. In addition, the blockchain keeps track of the AI's thinking processes as they take place. One must be able to analyse and comprehend information in order to be considered intellectually gifted. AI increases the intellect, understandability, clarity, and reliability of software by increasing its understanding, clarity, and dependability. AI has the potential to make the development of a blockchain a more efficient endeavour. Aims include improving security, reducing energy use, and increasing efficiency, among others [4].

Due in part to the fact that no studies have been conducted on AI and blockchain, there is still a great deal of material to study and comprehend. Even though there is integration between the two entities, it has not yet been systematically summarised and generalised, nor has a connection been established between the two. Additionally, a broad range of commercial sectors have seen significant development in recent years, with academia exploring the application of blockchain and AI [5]. This chapter investigates how blockchain and AI can work together in a comprehensive, multi-dimensional manner, conducting substantive research into a variety of areas in order to demonstrate how these concepts operate together in practice. Here is a brief description of the findings: This research made the following important contributions:

1. We're investigating the possibility of blockchain and AI working together, as well as the challenges associated with it.
2. We combine data from various categorisation schemes, which both local and international researchers have investigated, to provide a complete synopsis on the integration of blockchain and AI.
3. To help solve numerous real-world problems, we examine real-world application scenarios and practical use cases in a wide range of different areas.
4. We call attention to the difficulties that accompany blockchain and AI integration, and look forward to further study on these topics.

6.1.1 Normalisation Blockchain

In contrast to a conventional database, a blockchain with a network data structure is called a blockchain with network data structure rather than a conventional database. The Internet of Things Architecture and Technology is a novel parallel software architecture and technology that uses a centralised node acceptance mechanism to validate transaction information and coordinate the entire network, compared to previous technologies. The project is currently at an early stage of development, as stated. By using encryption technology, a data transmission can also be kept confidential and reliable. Figure 6.2 represents integration of blockchain system [6].

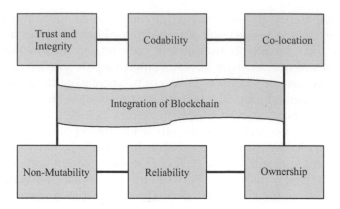

FIGURE 6.2
Illustration of blockchain characteristic with respect to integration.

6.2 Characteristics of Blockchain

6.2.1 Co-location

In order to enable the accounting, recording, storing, and updating of data without having a single central point, an E-chain blockchain makes use of distributed decentralised storage. Blockchain technology is decentralised, which enables any computer or node to participate, regardless of location.

6.2.2 Trust and Integrity

Each node on the network maintains its own ledger. In this case, the data being decrypted is not revealing any details about the people who are directly linked to it [7].

6.2.3 Ownership

The E-chain cryptocurrency blockchain was developed by third parties based on public rules and processes, and it is referred to as the E-chain system. Every node adheres to the applicable standards and procedures no matter what the system. By keeping complete control over every trustless transaction that is performed via our organisation, we ensure its accuracy and legality. E-chains with secure cryptographic components allow data transfers, recording, and updating, and actions that do not meet a standard are deemed ineffective.

6.2.4 Non-Mutability

When all network nodes have accepted the relevant data from the ledger and it has been committed to the blockchain, a local backup is performed. Using the hash mechanism, blocks are also correlated. You will need to change all the following blocks if you wish to manipulate data, which is an expensive endeavour.

6.2.5 Every Brick Stores the Entire History

Due to the logically centralised data containing all the cryptocurrency shop's data, every bit of data can be found simply by crossing the data from the web retailer.

6.2.6 Codability

In addition, it makes smart contracts more secure due to its decentralised nature. The result is an application ecosystem that is trustworthy for executing smart contracts. Contracts may be tailored by customers to suit their specific needs. Automating this process, letting you be certain that your assets and data are secure, it also protects your assets and data (Tables 6.1 and 6.2).

6.2.6.1 Information Transfer and Discovery

Application domains are currently addressing big data streams. Consolidated platforms for big data are also required [8]. The diversity of deep learning processes has increased dramatically. Knowledge management is based on the idea of providing personalised experiences. Knowledge trends serve the demands of all parties interested in computer hardware. Furthermore,

TABLE 6.1

Illustration of Blockchain Characteristic

S.No	Characteristic	B1	B2	B3	B4	B5	B6	B7	B8	B9
1	Co-location	✓	✗	✗	✓	✓	✗	✓	✓	✗
2	Reliability	✓	✓	✓	✓	✓	✓	✓	✓	✓
3	Efficient	✗	✓	✗	✓	✓	✗	✓	✓	✗
4	Systematic	✓	✓	✓	✓	✓	✓	✓	✓	✓
5	Duration	✓	✓	✗	✓	✓	✗	✓	✓	✗
6	Compatibility	✓	✓	✓	✓	✓	✗	✓	✓	✗
7	Reduction time	✓	✓	✓	✓	✓	✓	✓	✓	✓
8	System orientation	✓	✗	✓	✓	✓	✓	✓	✓	✓
9	Efficiency	✓	✗	✓	✓	✓	✗	✓	✓	✗
10	Intuitive	✓	✓	✓	✓	✓	✓	✓	✓	✓
11	Parallel computation	✓	✓	✓	✓	✓	✓	✓	✓	✓
12	Performance	✓	✓	✓	✓	✓	✓	✓	✓	✓

TABLE 6.2

Features of Blockchain in Sub Block Modelling

S.No	Feature	B1	B2	B3	B4	B5	B6	B7	B8	B9
1	Cooperation	✓	×	×	✓	✓	×	✓	✓	×
2	Dependability	✓	✓	✓	✓	✓	✓	✓	✓	✓
3	Accessed by	×	✓	×	✓	✓	×	✓	✓	×
4	Methodical	✓	✓	✓	✓	✓	✓	✓	✓	✓
5	Interruption	✓	✓	×	✓	✓	×	✓	✓	×
6	Adaptability	✓	✓	✓	✓	✓	×	✓	✓	×
7	Reduced time	✓	✓	✓	✓	✓	✓	✓	✓	✓
8	System-oriented	✓	×	✓	✓	✓	✓	✓	✓	✓
9	Amount of efficiency	✓	×	✓	✓	✓	×	✓	✓	×
10	Intelligent	✓	✓	✓	✓	✓	✓	✓	✓	✓
11	Multiprocessing	✓	✓	✓	✓	✓	✓	✓	✓	✓
12	Imaginative	✓	✓	✓	✓	✓	✓	✓	✓	✓

blockchain technologies have the potential to aid in the stable and traceable progression of ideas among AI apps that cover a wide range of themes.

AI and web crawlers acquire, define, and choose data on a daily basis through AI software and services, and then begin arranging it [9]. They use centralised preconceptions to extract information from the environment, techniques that result in amorphous data collecting. Data collecting from many views can be aided by centralised approaches. The money from the browser democratisation speeds up the process of tracing sensible pathways. Data encryption and storage, as well as indefinite storage systems as a result, approaches based on networked perception can be extremely beneficial. It is not necessary to obtain the content through the use of applications or networks sources of monetary gain as well as improved public perceptions. Because of the irreversibility of the blockchain, the only option is to complete the task at hand completely. It should have been possible to preserve the imprints of great perceptions. Cloud storage is a type of storage that is accessible from anywhere.

Automating and visualising data will require the use of neural networks in AI applications in the future. The concepts of ensemble, reinforcement, unsupervised, semi-supervised, and supervised are considered while analysing classification and regression problems. Deep learning techniques are also combined with notions of transfer. They can be used for a variety of machine learning tasks including classification and prediction, for example. Also, they can be employed for data analysis and clustering. The traditional learning models are explained and demonstrated in this section. A centralised system is employed in order to acquire a large amount of data. Local learning models can aid in the adoption of best practices in a variety of settings. This extends across all verticals in highly distributed and autonomous learning systems, allowing for fully coordinated local intelligence to be generated. Second, due to the fact that the blockchain preserves the origins and historical aspects of models, it allows for immutable and highly secure learning versioning to be implemented. In the event that smart data continues to grow indefinitely, contracts and learning models will have been thoroughly developed and tested. SEARCH AI applications should be capable of operating in both vast and sparse search contexts before being placed on the blockchain (i.e., large datasets or multivariable high-dimensional spaces with many variables). Because of this, methods for retrieving information are required for the most important information to be found [10].

Several factors will be considered during the design process, including completeness, complexity (time and space), and optimality. When it comes to nonlinear data structures, the vast majority of solutions are effective. Trees and graphs are examples of data structures where computers begin their work. Extraction of required variables or completion of traversal after expansion from a rough guide to the intended destination. The use of large-scale distributed infrastructure to conduct search solutions has become increasingly popular in order to improve operational efficiency. In contrast, the implementation of these technologies in a fragmented infrastructure will require significant investigation. It is intended that, in addition to standard search tactics, blockchains and other distributed database solutions will be utilised in the search process. It should be possible to see through it on a permanent basis, save fully rendered traces and navigation in a safe manner, and more. Avenues that can actually lead to optimal search solutions for other problems are being explored. According to current plans, the thought process will consist of systematic thinking as a critical component of AI systems because it enables programmers to construct inductive and deductive reasoning rules. The concept of "central thinking" is defined as follows: in the case of AI technologies, this results in more generic global behaviour, throughout the remainder of the programme. In order to address this issue, using blockchain-based dispersed reasoning techniques, it is possible to facilitate the development of customised reasoning systems with more perceptual-friendly methods of expressing oneself.

Also important is that the decentralised and distributed reasoning enabled by cryptocurrency smart contracts ensures that conceptual framework memories are available, which may be useful in performing similar reasoning in future strategies [11].

6.2.6.2 Computer Organisation

In addition to fast network storage, artificial capabilities are available. Information must always be designed in such a way that pertinent, correct, and full facts are retrieved via trusted websites supplies. High expressions are commonly used in mobile apps. Control schemes are also deployed throughout all nodes. In the blockchain network, these factors include the following: filters, categorisation, and dialogue files are now all forms of data management schemes. Considering limits on inefficient mutability and context networks and the use of blockchains makes data storage administration a tad easier. Because it will not only improve information duplication, but it will also be costly in the occurrence of simple data changes, even though it will be essential to repeat this export of similar datasets. For dealing with large amounts of data, this large data transmission will rapidly overwhelm resources and boost backhaul network traffic. As a result, autonomous data management has become critical for data AI. Redistribution at a certain point in the future, data management systems are planned to be implemented considering spatial and spatially node levels in the network the characteristics of the data additionally, global data analysis schemes can put metadata on the blockchain in a legal fashion in order to retain the data's security and origins. The actual information is stored on large-scale smart grids. Sensor networks and ensembles are illustrations of such systems. In this instance, metadata and actual data are used to detect consumer anomalies, are recorded on the ledger, and data is handled across a multiplatform. For sites that retain multiple shards or engage in swarms, the system uses gesture rewards. For sites that retain multiple shards or engage in swarms, the system uses gesture rewards [12].

6.2.6.3 Classification Algorithm Overview

In application domains, learning is essential to grasping the present state of the economy and make informed choices. Contemporary AI systems, on the other hand, need to minimise overfitting (i.e., producing exonerees on memorised data). Data that changes over time should be entered. Learning models, on the other hand, were created for consumer applications methods for decentralised learning pick appropriate bees and fragments for source sets and produces an extremely personalised classification model for each human clients. Also, blockchain technology has the potential to make transactions more efficient. Digital preservation and data integrity are also both critical factors. Studies have a strong record that may change over time. The author noticed autonomous learning methods are becoming more common. On the internet, this browser may become a model for someone else. Users do not even have to test the network from the beginning, just on this network. Rather than starting fresh, they can develop their own images in sections. New ones are acclimatised and educated. Unless you've found the procedure to be beneficial, several devices could have been used to test the learning approaches. In specialised, contracts play a vital role. Application of frameworks [4] a teaching model's genuine scheme is evaluated when it has been trained placement in real-world applications the model, nevertheless, Writers must upgrade and update models on a regular basis during deployment, which is both frequent and iterative inclinations (i.e., influencing a limited type of choices to be produced, neglecting all other possibilities) in order to generate incredibly effective results choice that is both productive and informative. These concerns are resolved using a striking approach that records information eternally. The modifications and the unchangeable version control several concepts. Moreover, model swapping between various companies is a worthwhile investment. And since, AI apps are more efficient and reliable. The programmers have a system to detect the origin and all the imprints of the solution, any version of a design in particular [17,18].

6.2.6.4 Uses of Cryptocurrency Forms for AI

Permissioned blockchain topology is a combination of bitcoin (in corporate, consortium, or surface settings, the blockchain apps only need to be authenticated) or a private blockchain (any client can find the blockchain apps in the public) system (web).

6.2.6.4.1 The Social Contract

Decentralised networks, including blockchain systems, are well known, whereby people can have the cryptocurrency code and embed it into their own devices system, adapt it, and employ it to fulfil their own goals. Additionally, blockchains are simple to use.

All participating nodes have access to data and can speak and understand it. User profiles can be accessed and are related to nature on major types, as well as transactions and sensitive data. Furthermore, anonymised and untraceable data was kept. Most blockchain systems make use of cutting-edge technology Public blockchains have grown in popularity since their widespread adoption. The players (i.e., consumers) on these blockchain systems are always private, despite their enormous complexity and transparency. As a result, hostile cybersecurity assaults on these bitcoins are always a threat, eventually leading to significant value and industrial espionage. Blockchains require a minimum majority of evaluators to achieve consensus or use algorithms to learn to crack access codes, due to high energy consumption, and are vulnerable to attack if attackers acquire control of 50% of the verifiers on the network. But compared to alternative and consortium blockchain systems, the transaction approval time on the blockchain is considerably longer. A blockchain transaction is typically completed in 10 minutes to a day, though this time is proportional to the number of network participants as well as the mathematical difficulty of the transaction. The intricacy of smart contracts is being utilised [13].

6.2.6.4.2 Personalised

A single entity manages a distributed ledger. For blockchain networks, distributed ledgers are typically developed as decentralised systems, with customers and parties well before memory (from transactions yet always available in the community). According to proven credentials, blockchain networks are significantly faster plasma, as well as coworkers who have never been to it before. As a consequence, more complicated math procedures are required. These networks are used to authenticate users. Additionally, anything in the form of local data, assets, or materials can be shared using a blockchain network. The platform's capabilities, among other things, were confirmed, and the transfer was successful. General elections or tripartite vehicle transactions are also used to handle commodity exchanges [14–15]. On either a cloud server, the time needed for an operation to be verified. The time it would take for a cryptocurrency to be constructed is generally as little as a millisecond [16].

6.3 Conclusion

The goal of this chapter was to show that various ideas and characteristics of blockchain technology may be used in a number of areas. Although these features have limited application in terms of currency and payments (Blockchain 1.0), and contracts, property, and all financial transactions (Blockchain 2.0), they have much broader implications that affect numerous other segments like the government, health, science, literacy, publishing, economic development, art, and culture (Blockchain 3.0). Blockchain technology has the potential to be very complimentary in a future world that incorporates both centralised and decentralised models, which is now being explored. As with any new technology, the blockchain is a concept that first disrupts, but over time may encourage the creation of a broader ecosystem that incorporates both the old and new ways of doing things. The invention of the radio caused sales of recorded music to skyrocket, while devices like the Kindle have boosted book sales. New York Times, blogs, Twitter, and customised drone feeds are all good sources of news. Media from both big entertainment corporations and YouTube are part of our media diet. As such, blockchain technology may exist in an ecosystem with both controlled and decentralised models, with time allowing for this to occur.

References

1. Maxmen, A., 2018. AI researchers embrace Bitcoin technology to share medical data. *Nature*, 555(7696), pp. 293–294.
2. Baynham-Herd, Z., 2017. Enlist blockchain to boost conservation. *Nature*, 548(7669), pp. 523–523.

3. Koch, M., 2018. Artificial intelligence is becoming natural. *Cell*, 173(3), pp. 531–533.
4. Deep Learning Neural Network Structure Optimization, 2020. Informatics and applications.
5. Calcaterra, C., Kaal, W. and Andrei, V., 2018. Semada technical whitepaper: Blockchain infrastructure for measuring domain specific reputation in autonomous decentralized and anonymous systems. *SSRN Electronic Journal*.
6. Dinh, T. and Thai, M., 2018. AI and blockchain: A disruptive integration. *Computer*, 51(9), pp. 48–53.
7. Qi, Y. and Xiao, J., 2018. Fintech. *Communications of the ACM*, 61(11), pp. 65–69.
8. Ozturan, C., 2020. Barter machine: An autonomous, distributed barter exchange on the ethereum blockchain. *Ledger*, 5.
9. Liu, B. and Ding, Z., 2021. A consensus-based decentralized training algorithm for deep neural networks with communication compression. *Neurocomputing*, 440, pp. 287–296.
10. Subramanian, H., 2019. Security tokens: Architecture, smart contract applications and illustrations using SAFE. *Managerial Finance*, 46(6), pp. 735–748.
11. Zheng, X., Zhu, M., Li, Q., Chen, C. and Tan, Y., 2019. FinBrain: When finance meets AI 2.0. *Frontiers of Information Technology & Electronic Engineering*, 20(7), pp. 914–924.
12. Baltrusaitis, T., Ahuja, C. and Morency, L., 2019. Multimodal machine learning: A survey and taxonomy. *IEEE Transactions on Pattern Analysis and Machine Intelligence*, 41(2), pp. 423–443.
13. Fioretto, F., Pontelli, E. and Yeoh, W., 2018. Distributed constraint optimization problems and applications: A survey. *Journal of Artificial Intelligence Research*, 61, pp. 623–698.
14. Ahmed, S. and Broek, N., 2017. Blockchain could boost food security. *Nature*, 550(7674), pp. 43–43.
15. Wright, C., 2008. Bitcoin: A peer-to-peer electronic cash system. *SSRN Electronic Journal*.
16. Portmann, E., 2018. Rezension, Blockchain: Blueprint for a new economy. *HMD Praxis der Wirtschaftsinformatik*, 55(6), pp. 1362–1364.

7

Adapting Blockchain for Energy Constrained IoT in Healthcare Environment

Ambika Nagaraj

St. Francis College

CONTENTS

7.1 Introduction

Internet of Things (IoT) (Ambika, 2019; Salman et al., 2018) is a prototype used in healthcare systems. The devices aim to analyze the captured inputs and transmit them by attaching their identifications. These devices include pervasive tiny devices and actuators providing readings to the network controller. These devices are low in storage. They assemble huge amount of data. These datasets are circulated with speed.

Different assessments can be performed by gathering execution measurements and using them with a proper energy model. They are some of the model requirements. It provides energy to other elements of the system. Three primary procedures gather execution measurements for energy utilization assessment. Register-Transfer Level (He et al., 2019) is difficult to accomplish fine-grained energy utilization attribution to the different programming parts. Also, demonstrating and profiling at such low levels are unfeasible for most business inserted processors since fundamental circuit data, like the powerful capacitance of major design blocks. Code instrumentation is performed by instrumenting the code with guidelines that remove execution penetrations. The challenge is to extricate the insights out of the equipment. It limits the overhead of the instrumentation that can essentially affect the assessment's exactness. Statistical PMC-based assessment system (Chadha et al., 2017) is intricate structures where ISA-level demonstrating also, investigations is inadequate to catch their intricacy. The insights build energy models and gauge the energy utilization of multi-strung/center structures. The power assessment can be empowered utilizing PMC. It takes into account energy-mindful choices.

The previous proposal (Feng et al., 2020) aims to minimize energy consumption in these nodes and expand the lifespan of them. The habiliment devices affixed to the patients gather enormous amount of data. The above layer consists of the fog devices that administer the data. A case study is considered in the proposal. A grouping framework is suggested. They group the devices and choose a principal node. This node is responsible to assemble the data of their respective cluster and transmit the processed one to the fog devices. To solve delay issue, these nodes are deployed at the boundary. This data can be retrieved by the authenticated user (doctors, personnel staff). The mist figuring climate is comprised of exceptional organization gadgets called haze hubs, which perform different processing assignments at the edge of the organization. Each mist hub can offer types of assistance, including preparing, memory, stockpiling, and organization transfer speed. In the lower layer, wearable sensors are conveyed on the patient's body to accumulate information. The wearable sensors screen and gather physiological information from patients as an EKG, blood oxygen, and other wellbeing-related data. The sensors conveyed assist patients with diminishing the burden of regular visits to specialists. Wearable sensors utilized in the lower layer may have restricted force, memory, preparation, and correspondence, so we will probably carry out a grouping strategy to amplify the organizational life of remote sensor organizations. The center layer comprises mist hubs. They prepare, store and transmit the information in the organization. As the medical care information gathered by the basic handheld gadget can expand, information mining and investigation of this enormous information are required. The mist hub in the center layer can deal with the crude information gathered in the lower layer for investigates.

The suggestion minimizes energy consumption by using hashing method. Every transaction transmitted for the first time is sent along with a hash code. If same transaction is to be sent again, only the hash code is sent. The work minimizes the energy consumption by 55.33% compared with the previous contribution. The work has five sections. Literature survey is briefed in Section 7.2. The proposed study is explained in Section 7.3. Analysis and simulation is described in the fourth section. The conclusion is summarized in the fifth section.

7.2 Literature Review

The section details various contributions done in this domain. The previous proposal (Feng et al., 2020) aims to minimize energy consumption in these nodes and expand the lifespan of them. The habiliment devices affixed to the patients gather enormous amount of data. The above layer consists of the fog devices that administer the data. A case study is considered in the proposal. A grouping framework is suggested. They group the devices and choose a principal node. This node is responsible to assemble the data of their respective cluster and transmit the processed one to the fog devices. To solve delay issue, these nodes are deployed at the boundary. This data can be retrieved by the authenticated user (doctors, personnel staff). The mist figuring climate is comprised of exceptional organization gadgets called haze hubs, which perform different processing assignments at the edge of the organization. Each mist hub can offer types of assistance, including preparing, memory, stockpiling, and organization transfer speed. In the lower layer, wearable sensors are conveyed on the patient's body to accumulate information. The wearable sensors screen and gather physiological information from patients as an EKG, blood oxygen, and other wellbeing-related data. The sensors conveyed assist patients with diminishing the burden of regular visits to specialists. Wearable sensors utilized in the lower layer may have restricted force, memory, preparation, and correspondence, so we will probably carry out a grouping strategy to amplify the organizational life of remote sensor organizations. The center layer comprises mist hubs. They are into preparing, stockpiling, memory, and organization transmission capacity abilities. As the medical care information gathered by the basic handheld gadget can expand, information mining and investigation of this enormous information are required. The mist hub in the center layer can deal with the crude information gathered in the lower layer for investigates.

It comprises three layers (Kaur & Sood, 2015) detecting and control layers, the data preparing layer, and the application layer. The SCL gathers information from the objective climate in an energy-productive way and communicates it to the IPL. The AL utilizes the data by extricating the IPL in different areas. Wellbeing checking, shrewd city, and brilliant transportation are some of the examples. The SCL comprises the components of an IoT framework (Ambika, 2020). It gathers crude information in huge volumes and sends them for information examination. The three principle segments of this layer are sensor hubs, energy-saving entryway hubs, and an energy-productive base station. The SNs are answerable for information assortment. They sense the objective climate and send the sensor estimations to a door hub. The AL offers types of assistance to the end clients. It is an interface to clients for applications. Some examples include wellbeing checking, keen city, keen transportation, climate observing, observation, business knowledge, savvy framework, and distant observing. The data from crude information gathered by the sensors are used by any application. The initial two sensors are occasional sensors. The third sensor is a trigger-based sensor that triggers when the HR increments over a specific breaking point.

The proposed framework (Al-Ali et al., 2017) is to screen and control the AC units; coordinated temperature and dampness sensor are interfaced with the microcontroller to quantify the encompassing conditions. It is constrained by the microcontoller to turn ON/OFF the gadgets likewise. A current sensor gauges the AC to figure the force consumption. In the proposed engineering, the workers are of good quality PCs. It conveys on cloud for wide-scale availability. The introduced workers are MQTT Broker, exceptionally adaptable Storage Server, Analytics Engine worker, and a web worker. Programming engineering comprises three essential structure modules: information securing module on the edge gadget, middleware module, and customer application module. The information module has two capacities, checking capacity and controlling capacity. The checking capacity constantly peruses the encompassing temperature and mugginess, and the AC power utilization communicates the readings to the middleware module through MQTT convention. These boundaries are outlined and answered to the middleware occasionally in standard MQTT design. For instance, the information outline has the client ID, house ID, gadget ID, and the sensor esteems. The control work gets the orders from the middleware module to turn ON/OFF the AC units in a like manner.

The proposal (Serra et al., 2014) is a unique energy scheduler with solace limitations, which considers both the keen estimating levies and the client's solace to choose the most energy-effective setup of HVACs that fulfills the client's necessities. We figure a streamlining issue of HVAC control by anticipating the temperature that a given arrangement of HVAC modules would cause in various areas. The usefulness and stream of data of the proposed engineering are clarified as follows. The temperature estimates areas through the WSN. Then, at that point, the estimations are intermittently shipped off the entryway, where the energy planning calculation is carried out. This calculation chooses the blend of the dynamic HVAC modules that limit the energy cost for given solace requirements and energy cost during a specific time frame. These choices are sent, through shell orders, to programmable flood defenders (actuators), which activate the HVAC modules. The HVAC modules change the room temperature as indicated by the choices taken by the energy scheduler.

During the observing stage (Moreno et al., 2014), data from heterogeneous sources is gathered and examined before proposing substantial activities to limit energy utilization thinking about the particular set of structures. Remembering that structures with various functionalities have distinctive energy use profiles, it is important to do an underlying portrayal of the principle supporters of their energy use. For example, in private structures, the energy is burned through, because the indoor administrations are with their tenants. In modern structures, energy utilization is related to the activity of mechanical apparatus and foundations committed to creation measures. The models anticipate the solace reaction of structure inhabitants given by the ASHRAE. A savvy executives framework should give appropriate transformation countermeasures to mechanized gadgets and clients, determined to fulfill the administrations of structures and energy productivity prerequisites. Robotization frameworks in structures take contributions from the sensors introduced in halls and rooms and utilize this information to control certain subsystems like HVAC, lighting, or security. These and more broadened administrations can be offered wisely to save energy, considering natural boundaries and the area of tenants. Input on utilization is vital for energy investment funds and ought to be utilized as a learning device.

The proposed framework model (Abedin et al., 2015) contains traditional gadget equipment. It addresses the physical equipment like sensors, home apparatuses. These gadgets associate with the implanted web worker. The inserted webserver has Restful web administration to speak with the cloud worker for the virtualization of objects. Virtual items will be moved to worker application for administration query and will have the virtualized object administration executable applications. The application worker interface will have the skeleton of the worker side code, and the customer won't get to the worker side code. The customer will speak with the application worker application through an interface. The administrator can get to the server application straightforwardly without the utilization of the worker interface. A bunch of low-controlled and less-proficient sensor devices is with IP address that goes about as transfer hubs. Then again, instruments with better availability and capacity go about as sink hubs that speak with the implanted gadgets. Transfer hubs create the information and send it to the sink hub so the sink hubs can divert the amassed stream of information to the installed web worker for additional handling. The proposed energy-proficient estimate is booking the obligation pattern of various sensors and machines. It outlines the proposed framework model. It can join with the planning prediction and can likewise fill its actual need. The calculation has three center stages, for example, on the job, Pre-off the clock, and Off-duty. In a working state, the gadget will perform with its full-fledged ability. Pre-off the clock state will actuate when the gadget is inactive for some time. The off-obligation stage holds three states to save energy in various conditions, and the energy effectiveness of the whole bunch or organization primarily relies upon this state. Sleep state is the state where the gadget will detect the climate before going into an energy-proficient state. No transmission or gathering of information will happen in this state. It can trigger the following energy-effective errand or move the condition of the gadget to a past state. Rest is a force-saving state that can rapidly permit the gadget to utilize the full-power activity. Force off is the most energy-proficient express that will place the gadget into profound rest. The utilization of energy should be zero since all kinds of energy devoured will halt at that point. The sink hub will trigger the gadget when any vital assignment ought to be performed.

The framework (Chooruang & Meekul, 2018) contains energy observing hubs that utilize the Peacefair PZEM-004T, minimal expense energy meter utilizing a noninvasive CT sensor, the SD3004 energy estimation chip, and microcontroller for estimating the voltage, current, dynamic force, and aggregate force utilization. The deliberate information will submit to the worker employing MQTT in JSON design. The Raspberry Pi 3 Model B runs as a neighborhood worker. Hence, clients admit to getting data about their energy utilization utilizing web applications locally or through the Internet. The PZEM-004T gives RMS voltage and RMS current and ascertains dynamic force and all-out energy utilization over the long run or collective force utilization. The PZEM-004T uses SD3004 energy estimation SoC chip from SDIC microelectronics. To send estimated information from the PZEM-004T to the organization or the web, we utilized the ESP8266 Wemos D1 smaller than usual to speak with the PZEM-004T through RS-232. The firmware for Wemos D1 which is smaller than expected was created using the Arduino programming conditions. The primary element is Wemos D1 small scale. It gathers energy information from the PZEM-004T and sends got information to the worker through Wi-Fi. The information ships off the worker roughly at regular intervals. The JSON design is a lightweight information exchange organization and straightforward. JSON design sends organized information over an organization association through MQTT.

7.3 Proposed Work

Table 7.1 is the representation of the notations used in the study. The previous proposal (Feng et al., 2020) aims to minimize energy consumption in these nodes and expand the lifespan of them. The habiliment devices affixed to the patients gather enormous amount of data. The above layer consists of the fog devices that administer the data. A case study is considered in the proposal. A grouping framework is suggested. They group the devices and choose a

TABLE 7.1

Notations Used in the Contribution

Notations	Description
C_i	i^{th} node of the cluster
H_i	Hash code generated at i^{th} time
D_i	Data set
C_{id}	Identification of the node
S_i	Server

principal node. This node is responsible to assemble the data of their respective cluster and transmit the processed one to the fog devices. To solve delay issue, these nodes are deployed at the boundary. This data can be retrieved by the authenticated user (doctors, personnel staff). The mist figuring climate is comprised of exceptional organization gadgets called haze hubs, which perform different processing assignments at the edge of the organization. Each mist hub can offer types of assistance, including preparing, memory, stockpiling, and organization transfer speed. In the lower layer, wearable sensors are conveyed on the patient's body to accumulate information. The wearable sensors screen and gather physiological information from patients as an EKG, blood oxygen, and other wellbeing-related data. The sensors conveyed assist patients with diminishing the burden of regular visits to specialists. Wearable sensors utilized in the lower layer may have restricted force, memory, preparation, and correspondence, so we will probably carry out a grouping strategy to amplify the organizational life of remote sensor organizations. The center layer comprises mist hubs. They are into preparing, stockpiling, memory, and organization transmission capacity abilities. As the medical care information gathered by the basic handheld gadget can expand, information mining and investigation of this enormous information are required. The mist hub in the center layer can deal with the crude information gathered in the lower layer for investigates.

The suggestion aims in minimizing the energy consumption in the system. Hence hashing methodology is adopted. The wearables (Nagaraj, 2021) gather information from the sick and communicate to the the same to the pre-defined destination. The reading would be the same most of the times with an exception during the critical times. The proposal adopts clustering method, where the group head gathers reading from other detecting devices and dispatches the processed data to the server. The regular readings can be conveyed in the compressed manner.

7.3.1 Generating Hash Code

After the cluster head gathers the data, it analyzes the data and generates the hash code for the data gathered. This code is attached to the data and dispatched to the server. In the equation (7.1) the cluster head C_i is attaching the data set D_i, hash code H_i, and its identification C_{id} and transmitting to the server S_i.

$$C_i : D_i \| H_i \| C_{id} \to S_i \tag{7.1}$$

Every time if there is variation in data, similar method is adopted. When the head encounters same data sets at different intervals of time, it dispatches only the hash code along its identification. In equation (7.2), the cluster head C_i is transmitting the hash code H_i and its identification C_{id} to the server S_i.

Hence, a large amount of energy is saved. When the new cluster head is elected, the previous head forwards the data sets and related hash code to it. Table 7.3 is the representation of hash code generation.

$$C_i : H_i \| C_{id} \to S_i \qquad (7.2)$$

7.4 Analysis and Simulation

The work is simulated in NS2. Table 7.2 contains the description of the simulation work. Table 7.3 lists the steps of the generation of hash code algorithm.

7.4.1 Energy Consumption

Energy is one of important ingredients for the sensors. These devices are not recharged. Hence conserving energy is one such measure to keep it alive for a long time. One of the methods adopted similar to the previous contribution

TABLE 7.2

Parameters Used in Simulation Work

Parameter	Narration
Dimension of the network	$200\,\text{m} \times 200\,\text{m}$
Number of devices under consideration	40
Distribution of the sensors	Manual (uniform)
Number of clusters formed	8
Length of data considered	256 bits
Hash code generated	10 bits
Identification length	24 bits
Simulation time	60 ms

TABLE 7.3

Generation of Hash Code

Step 1: Input the aggregated data (256 bits)

Step 2: Divide the data into two sets (128 bits)

Step 3: Divide the first set by 64 (in binary form) and concatenate the remainder bits and the quotient aside (2 bits +3 bits)

Step 4: Divide the second set by 32 (in binary form) and concatenate the remainder bits and the quotient aside (2 bits +3 bits)

Step 5: Concatenate both of them

Step 6: Apply right circular shift 2 bits

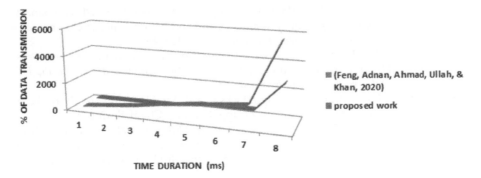

FIGURE 7.1
Comparison of energy consumption in both the systems.

is clustering. The suggestion also adopts using hashing to save energy. It conserves 53.33% of more energy compared with previous work (Feng et al., 2020). Figure 7.1 represents the same.

7.5 Conclusion

IoT devices are the assembly of many sensors and actuators with intelligence. These devices aim to minimize human efforts. They are unsupervised system capable of accomplishing the tasks without human intervention. The previous work aims to minimize energy consumption in healthcare system by filtering message. The habiliment devices affixed to the patients gather an enormous amount of data. The above layer consists of the fog devices that administer the data. A case study is considered in the proposal. A grouping framework is suggested. They group the devices and choose a principal node. This node is responsible to assemble the data of their respective cluster and transmit the processed one to the fog devices. To solve delay issue, these nodes are deployed at the boundary. This data can be retrieved by the authenticated user. The suggestion aims in minimizing the energy consumption by using hashing. The energy is conserved by 55.33% compared to the previous contribution.

References

Abedin, S. F., Alam, M. G., Haw, R., & Hong, C. S. (2015). A system model for energy efficient green-IoT network. *International Conference on Information Networking (ICOIN)* (pp. 177–182). Cambodia: IEEE.

Al-Ali, A.-R., Zualkernan, I. A., Rashid, M., Gupta, R., & Alikarar, M. (2017). A smart home energy management system using IoT and big data analytics approach. *IEEE Transactions on Consumer Electronics, 63*(4), 426–434.

Ambika, N. (2019). Energy-perceptive authentication in virtual private networks using GPS data. In: Mahmood, Z. (ed.), *Security, Privacy and Trust in the IoT Environment* (pp. 25–38). Cham: Springer.

Ambika, N. (2020). Tackling jamming attacks in IoT. In: Alam, M., Shakil, K. A., & Khan, S. (eds.), *Internet of Things (IoT)* (pp. 153–165). Cham: Springer.

Chadha, M., Ilsche, T., Bielert, M., & Nagel, W. E. (2017). A statistical approach to power estimation for x86 processors. *IEEE International Parallel and Distributed Processing Symposium Workshops (IPDPSW)* (pp. 1012–1019). Lake Buena Vista, FL: IEEE.

Chooruang, K., & Meekul, K. (2018). Design of an IoT energy monitoring system. *6th International Conference on ICT and Knowledge Engineering (ICT&KE)* (pp. 1–4). Bangkok, Thailand: IEEE.

Feng, C., Adnan, M., Ahmad, A., Ullah, A., & Khan, H. U. (2020). Towards energy-efficient framework for IoT big data healthcare solutions. *Scientific Programming, Hindawi, 2022,* 1–9.

He, M., Park, J., Nahiyan, A., Vassilev, A., Jin, Y., & Tehranipoor, M. (2019). RTL-PSC: automated power side-channel leakage assessment at register-transfer level. *IEEE 37th VLSI Test Symposium (VTS)* (pp. 1–6). Monterey, CA: IEEE.

Kaur, N., & Sood, S. K. (2015). An energy-efficient architecture for the Internet of Things (IoT). *IEEE Systems Journal, 11*(2), 796–805.

Moreno, M., Úbeda, B., Skarmeta, A. F., & Zamora, M. A. (2014). How can we tackle energy efficiency in IoT basedsmart buildings? *Sensors, 6,* 9582–9614.

Nagaraj, A. (2021). *Introduction to Sensors in IoT and Cloud Computing Applications.* Dubai, UAE: Bentham Science Publishers.

Salman, O., Elhajj, I., Chehab, A., & Kayssi, A. (2018). IoT survey: An SDN and fog computing perspective. *Computer Networks, 143,* 221–246.

Serra, J., Pubill, D., Antonopoulos, A., & Verikoukis, C. (2014). Smart HVAC control in IoT: Energy consumption minimization with user comfort constraints. *The Scientific World Journal,* 1–12.

8

Security and Privacy Issues in IoT Systems Using Blockchain

Samarth Vashisht
Synopsys, Inc.

Smriti Gaba
Reliance Jio Infocomm Limited

Susheela Dahiya and Keshav Kaushik
University of Petroleum & Energy Studies (UPES)

CONTENTS

DOI: 10.1201/9781003193425-8

8.1 Introduction

IoT (Internet of Things) is a fast emerging technology in the mainstream industry. It is connecting people, places, and products, in the process of which value creation is expanding. On the core side, IoT systems contain chips, sensors, and other components that transmit data to the IoT network, and this system being "Smart" comes with the capabilities of analysing transmitted data, converts these insights into implementations, and furthermore, induces new methods of connectivity, growth along with business and industry-wide opportunities (Domagala, 2019). IoT being a revolutionary technology with massive connectivity of device-to-device communication has improved many sectors in industry including healthcare, factories, traffic network, Smart homes, etc.; however, it brings forth many security challenges including authentication, access control, privacy etc. As growth of these devices is increasing exponentially, the challenge of secure identification of nodes, further authentication process, and other security aspects also increases. For such resource-constrained systems and networks, it is quite a difficult task to implement a Centralised Security System as not only the single point of failure will lead to Distributed Denial-of-Service attack (DDoS) attack, but it would also be a challenge to Centralise Industrial Infrastructure as edge devices in industrial setup may exist in widespread geographical locations.

IoT systems and services are associated with critical data, industrial infrastructure networks, and assets which make them vulnerable and thus prone to attacks and data breaches. Considering the fact that IoT systems are resource-constrained, there arise limitations to deploy conventional security solutions. As an IoT system or network involves decentralised architectures, a security solution that comes with a decentralised model has to be implemented with limited resources. In such a scenario Blockchain being a decentralised approach to security can be deployed with these IoT systems to maintain security and privacy along with increasing the scalability of IoT.

Blockchain also termed as distributed ledger technology is one of the most reliable security solutions recommended for IoT systems (Huang et al., 2019) as it provides protection to IoT applications by blocking any data tampering or manipulation of records, identifying compromised nodes, and removing them from the network to maintain confidentiality, integrity, and availability of the complete IoT system.

Blockchain is a distributed technology that provides a collection of records where transactions and data can be stored. These records are immutable which further makes tampering of data harder, thus improving security by sending the records over a distributed network to the involved nodes. During the process, Blockchain approach uses strong cryptographic functions and public keys for encryption, maintaining strong hashes using cryptographic techniques ensuring integrity, and moreover, reducing risks of data tampering during any transaction in the communication.

8.2 Use of Blockchain in IoT

Blockchain acts as the missing link between Security and IoT. In the blockchain technology, blocks are considered as key aspects that basically are the transactions or sets of transactions occurring within the system. Blocks work based on the previous reference approach, and each block when introduced in the system contains a reference to the preceding transaction along with a strong cryptographic hash of the transaction. This way a "Chain" of blocks is created, and therefore, it is called Blockchain. These blocks are created by "Miners". In the network based on a peer-to-peer model, any of the nodes can choose to be a miner that uses powerful processors to mine blocks by solving cryptographic functions and computations called Proof-of-Work (Liu et al., 2019) and then appending mined blocks to the blockchain. As soon as a new transaction takes place within the system and is broadcasted over the network, miners start the verification process of the new transaction, validate it, and each one of them chains the transaction to its own pending block of yet-to-be mined transactions. This way a single transaction is being processed by multiple nodes, thus increasing the robustness of the Blockchain system, and moreover, making it scalable and reducing the risks of single point failure (Agarwal & Pal, 2020).

This concept of blocks is the reason behind tamper-resistant ability of blockchain-based security solutions as creating these blocks requires difficult computations along with specialised resources and time for creation, and therefore, to tamper any of the blocks, the previous one has to be tampered, that is again chain references to its preceding block and the chain goes on, thus making it difficult to compute and hence tamper-proof.

A secure mesh network can be created using blockchain technology and its concept; within this secure network of IoT devices, nodes securely connect to each other for communication, avoiding possible cases of spoofing or impersonation attacks. This can further improve security as each node in the network will be registered with uniquely identifiable blockchain ID, such that whenever a node tries to connect to another node for device-to-device communication, it can use the Blockchain ID as a URL along with its local blockchain wallet for identification. The blockchain wallet generally creates a digitally signed request that could be used for validation by other device using public key of the requester and hence secure authentication can be achieved with the benefits of transparency to the authorised nodes enabling them to identify the reliability of other connected nodes as well as view past transaction records further providing an efficient mechanism to identify the source of any data leakages or data breaches within the system, take appropriate remedial actions to avoid risks and reduce the attack surface of IoT system. The number of IoT-driven systems is increasing exponentially with increase in interconnectivity among these devices and networks causing delays in processing, storing large amounts of data. Blockchain supports these systems by its distributed ledger technology to enable fast processing of transactions.

Advantages of using Blockchain in IoT systems can be summarised as below:

i. **Decentralisation**: The decentralised framework provided by block-chain reduces the chances of single point failure (Yu et al., 2018), thus improving the efficiency and availability of the IoT system as even if one or more nodes go offline the availability of the peer-to-peer network is not compromised. The data is being stored in multiple nodes on the network making it resistant to attacks and failures.

ii. **Secure Transactions**: Blockchain uses strong cryptographic techniques to encrypt the transactions, link them to the previous transaction to enhance ability of resistance toward data tampering, storing the transactions, and data across distributed network which secures the data from attackers, using Public/Private key mechanism with asymmetrical cryptographic techniques which generates the keys randomly going through a number of computations that are hard to formulate, thus providing protection against attacks. It further provides a mechanism for verifying and validating the nodes which ensures a secure way of communication, correctness of data, and immutability among the devices in the system.

iii. **Unique Identification**: In the IoT network, each node or device involved is assigned a blockchain ID that provides unique identification of each device and thus ensures a trusted distributed platform for storing data with security.

iv. **Traceability**: Blockchain system maintains records of past transactions, which allows tracking of any record or checking authenticity of assets to avoid or detect any fraud, impersonation attacks. This functionality of Blockchain systems can be used across IoT systems involving supply chain of goods to ease tracking of records.

v. **Reliability**: Any transaction before entering the Blockchain system is verified by the miners. Therefore, all nodes in the Blockchain IoT system can authenticate the information that is being passed in the network as the data is reliable and verified beforehand.

vi. **Data Privacy**: One of the security concerns in resource-constrained IoT systems is protection of Data Privacy. With utilisation of Blockchain Technology in IoT systems, its features including the immutability and reliability of data enhance Data Privacy protection along with keeping the records of all transactions in archive that can easily be monitored by the authenticated nodes present in the network which retains the data rights and privacy policies.

8.3 Security and Privacy Issues

IoT is emerging and generating vast amounts of data. It is being adopted across all sectors including healthcare, supply chain, industrial, agriculture, and other infrastructure businesses. An IoT system contains sensors and other components which communicate with each other to exchange data, processing, and other computational operations. Despite having numerous advantages, IoT faces many security and privacy challenges.

8.3.1 Security Issues

8.3.1.1 Authentication

Authentication is a critical security issue in IoT systems as the implementation has applications in healthcare, vehicular, network critical infrastructure. Therefore weak authentication may allow attackers to gain access to the device, which could further lead to data tampering of real-time data, stealing of sensitive information thus leading to compromise of confidentiality, integrity, and availability.

IoT devices have limited access to resources; therefore, they are incapable of handling cryptographic calculations, due to which they are not able to provide optimised and better mechanisms for authentication. In such cases, attackers may bypass authentication, gain access to devices, and modify parameters, further compromising the IoT system.
Countermeasures:

- Implementing strong cryptographic techniques such as Public Key Infrastructure
- Utilisation of Blockchain-based systems integrated with Cloud storage which implements proxy encryption scheme, smart contracts (Bodkhe et al., 2020) to ensure only authorised people get access to the system and identify the authenticity of users.

8.3.1.2 Network Security

Many IoT systems include components that provide wireless networking and services, introduce security issues, and increase potential attack surface. These wireless devices connected in the network transmit private data, and if this communication is not secure, it may lead to sniffing attacks and leakage of private data.

Countermeasures:
Distributed blockchain-based systems can provide encrypted and secure communication by linking new transactions to previous transaction blocks, forming a chain of immutable (Tanwar et al., 2020) blocks further making it difficult to sniff and tamper data.

8.3.1.3 Access Control

To secure authority of users, access control techniques must be implemented. In case no access level and policies are applied, it may allow unauthorised access to resources, data in an IoT system leading to fraudulent, malicious activities.

Countermeasures:
Blockchain-based smart contracts and access control list can be implemented to ensure no unauthorised node is introduced in the system.

8.3.2 Privacy Issues

Nodes participating in IoT networks collect a lot of sensitive information from the end-user that raises privacy concerns. To preserve privacy in IoT systems, blockchain-based secure techniques are implemented that ensure integrity and confidentiality of data, and further provide protection against data tampering.

8.3.2.1 Data Privacy

IoT devices are resource-constrained due to which they may not use strong cryptographic techniques to encrypt or decrypt data being generated or processed. If data is transmitted unencrypted, it can be obtained by attackers compromising the privacy of the data.

To preserve privacy of data, strong cryptographic techniques and use of randomly generated public/private keys must be used by deploying blockchain models so that data is transmitted encrypted.

8.3.2.2 Usage Privacy

The usage pattern of users using IoT applications and services incorporates privacy issues that need to be protected. For example, the data collected or generated by smart video surveillance reveals private information of households and the places where it is deployed. Similarly, privacy of healthcare systems must be preserved as a data breach in this case may result in sharing of critical information and private data. Implementation of privacy-preserving techniques is crucial for enhancing security.

8.3.2.3 Location Privacy

Many IoT systems involving wireless sensor applications provide services based on location. This location-based information when generated and transmitted among the network raises privacy issues. Therefore, appropriate techniques must be deployed to protect confidentiality of the data being shared.

8.4 Blockchain for Securing IoT Systems

Blockchain provides robustness, faster processing, and reliable security solutions that can be deployed in IoT systems to overcome security and privacy challenges in different IoT applications. The areas where Blockchain can be used as a solution for securing IoT systems are discussed below.

8.4.1 IoT eHealth

IoT eHealth provides hospitals a platform that gives rapid access to medical services and the healthcare service system of their hospital. IoT devices and its components are widely used in medical equipment and healthcare sectors for such services. However, the use of IoT devices and sensors also introduces security and privacy issues to the healthcare system as well as to patients' medical records. The adoption of Blockchain for eHealth can address the security and privacy issues of the framework by providing secure authorisation and integrity of medical records, securely tracking past records and prescriptions to avoid any repetition of medical tests, protection against other fraudulent activities, and data tampering attacks.

Blockchain Solutions for eHealth:

1. An effective Blockchain-based monitoring platform (Jamil et al., 2020) contains a Blockchain network with authorised nodes, the patients with the sensor wearables in case of any emergency can send signs through their node to the peer-to-peer network, and the authorised nodes on this network (hospital facilities) can monitor these signs and take appropriate action by identifying the source of transactions of the vital signs along with his details and records.

   Security and Privacy: During the implementation of the platform, Smart contract is deployed which ensures that the vital sign transactions on the Blockchain network are encrypted and confidential. Along with this to ensure that only authorised users access this information over the network, access control policies are implemented to maintain privacy of the data.

2. To preserve the privacy of data being exchanged over the healthcare system (Kaur et al., 2018), cloud storage is integrated with the IoT system such that the storage of electronic medical records is carried out securely by utilising the Smart contract technology that regulated encryption of the medical records being stored and blockchain that further maintains records of these transactions for effective, secure traceability.

3. Access control in eHealth services is one of the key aspects that needs to be implemented securely. Blockchain-based systems store access policies and maintain transaction verification that ensures the medical records, health services are only accessible to authorised users with appropriate privileges. To maintain the privacy of data, many blockchain-based frameworks are introduced that are effective in securing the eHealth IoT system (Tanwar et al., 2020).

4. IoT systems are resource-constrained, while a Blockchain-based solution requires high computation power and resources for implementation of cryptographic mechanisms, which the IoT devices are not able to provide. Therefore, researchers (Ismail et al., 2019) proposed a lightweight blockchain solution for IoT devices where a head blockchain manager is responsible for managing the transactions and creating blocks instead of multiple nodes, which further avoids forking of any transaction on other nodes thus reducing computational costs.

8.4.2 Smart Cities and Smart Home

The convergence of fog Computing and Cloud Computing with IoT Systems resulted in the evolution of Smart Homes and Smart cities. These come with a range of different components which include IoT devices, sensors, networks, data storage, processing and computation of the data utilising storage of cloud, features of Fog computing for processing, and analysis of data to avoid any delays resulting in an efficient Smart system with high availability and performance. However, the convergence of these technologies brings multiple security and privacy issues and therefore needs a robust mechanism to overcome these challenges that could provide protection against fraudulent activities and attacks. Blockchain could be one of the solutions that can be deployed to build a secure and effective system.

Current Smart cities/Smart Home solutions deployed need different services like storage, processing, and analysis of data. For these services, Smart Homes widely depend on third-party services which include cloud for storage and Fog models for processing and analysis, which are responsible for managing the data during the network communication between components of smart homes, resulting in vulnerabilities and increased probability of single point failure (Botta et al., 2015). Apart from dependency, these Smart Home systems capable of providing faster communication are centralised leading to scalability issues. This issue is also addressed by the decentralised, distributed ledger technology Blockchain providing high scalability, reliability, and secure communication between components.

Blockchain Solutions for Smart Cities and Smart Home:

a. Blockchain-based systems along with Smart contract can be used in a Smart city infrastructure in which all the transactions from IoT devices are securely stored in Cloud storage ensuring authenticity, non-repudiation, and confidentiality of the data and communication between the components (Sookhak et al., 2019).

b. Implementing Blockchain-based solutions in this interconnected smart city network gives rise to instability of blockchain mechanism due to incapability of IoT devices to handle high power cryptographic computations. Therefore, Lightweight Blockchain systems can be implemented to reduce computational costs and improve efficiency of Smart city infrastructure (Paul et al., 2018).

c. In a Smart Home network, there are many components including IoT devices, hubs, sensors, and motion detectors that collect data based on the environment, for example, temperature sensors capture room temperature, and other information which needs to be stored on a server or say a cloud storage platform. The communication that takes place during this process is vulnerable to attacks and privacy concerns. In this scenario, if Blockchain is integrated with cloud technology, it forms a distributed and secure framework for protection and integrity of the data being processed and stored over the network (Abu Naser & Alkhatib, 2019).

8.4.3 IoT Vehicular Network

IoT systems expand its growth to evolve Smart transportation systems involving integration with other technologies. The shortcomings of this transportation can be overcome using the storage capabilities of Cloud technology, the latency issues along with better processing can be achieved by integrating Fog computing model, and to add more security, high availability, reliable intelligent transportation system, Blockchain technology is combined with these technologies. The integration of these technologies gives rise to a secure and efficient vehicular network system.

Blockchain Solutions for Vehicular Networks:

a. A blockchain-based solution along with integration of cloud technology can be implemented to provide an ad-hoc distributed vehicular network (Nadeem et al., 2019) where the storage and computation issues are addressed by the Cloud architecture and the peer-to-peer network of Blockchain technology further enhances the security and efficiency of the system protecting it against cyber-attacks.

b. The intelligence of software-defined networking (SDN) with its controlled management when integrated with Fog computing model results in a high-performance extended application of SDN enabled 5G vehicular network (Xie et al., 2019). These systems include an architecture such that each vehicle can share road information contained in a tag with other connected vehicles in the network. Blockchain technology is integrated in this model to improve the authenticity of vehicles involved in the network along with the records of past information to identify the source of information reliably.

c. Self-driving vehicles is a use case scenario where humans will rely completely on the vehicles for services. Therefore it is essential that all the services and communication between them are efficient. In this case, communication will take place for information between committed networks of vehicles; therefore, to secure this system and maintain the records of transactions, Blockchain technology could be a perfect fit providing security, efficiency, and scalability of the driverless vehicular network (Pedrosa & Pau, 2018).

8.4.4 Software-Defined Networks

The emergence of SDN enabled organisations to have customised network services which can be controlled centrally using software applications. This incorporates many security challenges in the IoT network along with huge traffic and bandwidth issues. Therefore, there is a need to deploy Blockchain-based distributed technology to enhance security of IoT networking.
Blockchain Solutions for SDN:

a. Blockchain along with integration of a proxy encryption technique (Gao et al., 2020) can be used to improve the security and authenticity of software-defined networking. Smart contracts are utilised in order to keep track of records in this solution.

b. SDN provides flexibility to the users of the network which increases security threats including unavailability of information related to routing and forwarding decisions among the devices in the network that further leads to failure in communicating the SDN controller introducing many vulnerabilities into the system. Therefore, Blockchain is deployed in such cases to provide a network monitoring aspect where SDN controllers can keep track of records and monitor them for information (Misra et al., 2020).

c. Various security attacks are common in software-defined networks; for example, if the SDN controller is attacked, it may forward malicious or fraudulent information or may lead to modification of instructions, policies, and rules which could affect the complete network. In this scenario, features of blockchain technology could be utilised to keep track of SDN controller records, logs, and events in the network to detect any suspicious activity (Zhang et al., 2020).

d. DDoS attacks in SDN can be prevented by deploying blockchain-based DDoS detection (Abou El Houda et al., 2019) and prevention mechanisms that use smart contracts to distribute any attack information among the SDN-connected domains so that other connections can identify and collaborate to take appropriate actions to improve security of the network.

8.4.5 Wireless Sensor Networks

Wireless sensor networks are a part of many IoT systems that collect information and perform remote sensing operations for applications involved. For example, environmental sensing applications require temperature sensors, pollution sensors, and pressure sensors that require collection of information readings and performing computations on them. In such scenarios, nodes in the IoT network remain unattended which introduces vulnerabilities in the system and failure of operations.

Blockchain Solutions for Wireless Sensor Networks:

A blockchain-based solution could be deployed to identify the failed or inactive node by keeping record of status of activity for all nodes. In case of any failure node, the cluster head responsible for maintaining the records of blockchain utilises the smart recovery contract to record status of all active nodes and detect the failure node.

8.4.6 Supply Chain and IoT

IoT-based frameworks for supply chain management systems are improving business sectors, as they contain sensors and devices with GPS modules to track and maintain records of shipment status of products, authentication management, and monitoring if the storage of products is in appropriate conditions. However, they introduce an attack surface and are prone to many vulnerabilities.

Blockchain Solution for Supply Chain:

Blockchain-based framework can be integrated in the architecture of IoT-based supply chain management systems, through which the records of shipments can be easily tracked and managed securely with efficient cryptographic encryption techniques. Access Control List (Malik et al., 2018) can be included in the system to manage access of authorities for reading and writing of transactions along with protection against any data tampering during the process.

8.5 Challenges in Adopting Blockchain for IoT System

1. IoT systems have limitations on power consumption and resources, while Blockchain implementation requires high consumption and powerful resources to carry out excessive mathematical computations along with storage capacities. In case blockchain is deployed on any IoT-based system, it would require capacity to mine blocks for a stable peer-to-peer communication network with efficiency.

   Possible Solution: However, there are some methods that can be utilised in order to resolve this issue that includes the patient-centric agent (Uddin et al., 2018) approach which runs on cloud and edge server, performs required operations, and handles multiple blocks on these servers so that blockchain can be accommodated for IoT systems. This PCA software runs on a device or node which if gets stolen may lead to compromise of security; therefore, this solution may not be a reliable one in terms of security of the system.

2. As blockchain is based on data distribution, it broadcasts the new blocks and transactions among all the participants of the peer-to-peer network which needs higher bandwidth support. However, IoT devices are resource-constrained and thus have limited bandwidth; therefore, they cannot handle the consumption of bandwidth by integrated blockchain mode further causing unstable blockchain mechanisms. In this scenario, patient-centric agent mechanism can be utilised to reduce bandwidth requirements in blockchain network by only allowing significant transactions in blockchain network.

3. IoT devices generate large amounts of data and stream it continuously on the network. Storing and processing all the data on a chain is one challenging task for Blockchain technology. To overcome this challenge, research suggested off-chain (Uddin et al., 2020) methodology in which IoT data is distributed among different repositories; for storage of data, Cloud repositories are utilised;

and for processing and computation of data repositories on local network, blockchain networks are utilised as per the context and structure of the data.

4. In blockchain networks, the feature of transparency among the nodes connected in the network leads to privacy issues, as all nodes can access the information that is being passed in the network. This gives rise to data leakage issues as well as privacy concerns. Therefore, blockchain networks sometimes utilised the implementation of homomorphic encryption (Shrestha & Kim, 2019) techniques to ensure security of data and privacy of miners as if in case any third-party services try to access data for performing any computation or operations, then homomorphic encryption allows these operations on the ciphertext without need of decryption of data, thus preserving privacy.

8.6 Conclusion

Blockchain is an emerging technology used in many sectors to ensure security and privacy including bitcoins and cryptocurrencies. The adoption of such extensive technology for IoT systems can provide a more efficient and distributed architecture to the IoT world. The purpose of this study is to review the current state of IoT applications and the optimisations, transformations that can be incorporated to improve the IoT sector by integration of different technologies including Cloud Computing for storage, Fog computing for better processing, and Blockchain for ensuring strong cryptographic techniques are implemented to secure transactions, maintaining the privacy of data involved in IoT applications. However, despite incorporation of different technologies, many security and privacy issues remain unaddressed in IoT systems while also introducing new challenges related to bandwidth and connectivity of the network. Furthermore, new research and solutions are also discussed that can be implemented to overcome these challenges, preserving the efficiency of IoT systems.

References

Abou El Houda, Z., Hafid, A., & Khoukhi, L. (2019). Co-IoT: A collaborative DDoS mitigation scheme in IoT environment based on blockchain using SDN. *In 2019 IEEE Global Communications Conference (GLOBECOM)*, Waikoloa, HI, (pp. 1–6). IEEE.

Abu Naser, M., & Alkhatib, A. A. (2019). Advanced survey of blockchain for the internet of things smart home. *In 2019 IEEE Jordan International Joint Conference on Electrical Engineering and Information Technology (JEEIT),* Amman (pp. 58–62). IEEE.

Agarwal, V., & Pal, S. (2020). Blockchain meets IoT: A scalable architecture for security and maintenance. *In 2020 IEEE 17th International Conference on Mobile Ad Hoc and Sensor Systems (MASS),* Delhi (pp. 53–61). IEEE.

Bodkhe, U., Tanwar, S., Parekh, K., Khanpara, P., Tyagi, S., Kumar, N., & Alazab, M. (2020). Blockchain for industry 4.0: A comprehensive review. *IEEE Access,* 8, 79764–79800.

Botta, A., De Donato, W., Persico, V., & Pescapé, A. (2016). Integration of cloud computing and internet of things: A survey. *Future Generation Computer Systems,* 56, 684–700.

Domagala, P. (2019). Internet of Things and Big Data technologises as an opportunity for organizations based on Knowledge Management. *In 2019 IEEE 10th International Conference on Mechanical and Intelligent Manufacturing Technologies (ICMIMT),* Cape Town (pp. 199–203). IEEE.

Gao, Y., Chen, Y., Lin, H., & Rodrigues, J. J. (2020). Blockchain based secure IoT data sharing framework for SDN-enabled smart communities. *In IEEE INFOCOM 2020-IEEE Conference on Computer Communications Workshops (INFOCOM WKSHPS),* Toronto, ON (pp. 514–519). IEEE.

Huang, J., Kong, L., Chen, G., Wu, M. Y., Liu, X., & Zeng, P. (2019). Towards secure industrial IoT: Blockchain system with credit-based consensus mechanism. *IEEE Transactions on Industrial Informatics,* 15(6), 3680–3689.

Ismail, L., Materwala, H., & Zeadally, S. (2019). Lightweight blockchain for healthcare. *IEEE Access,* 7, 149935–149951.

Jamil, F., Ahmad, S., Iqbal, N., & Kim, D. H. (2020). Towards a remote monitoring of patient vital signs based on IoT-based blockchain integrity management platforms in smart hospitals. *Sensors,* 20(8), 2195.

Kaur, H., Alam, M. A., Jameel, R., Mourya, A. K., & Chang, V. (2018). A proposed solution and future direction for blockchain-based heterogeneous medicare data in cloud environment. *Journal of Medical Systems,* 42(8), 1–11.

Liu, Y., Wang, K., Lin, Y., & Xu, W. (2019). LightChain: A lightweight blockchain system for industrial internet of things. *IEEE Transactions on Industrial Informatics,* 15(6), 3571–3581.

Malik, S., Kanhere, S. S., & Jurdak, R. (2018). Productchain: Scalable blockchain framework to support provenance in supply chains. *In 2018 IEEE 17th International Symposium on Network Computing and Applications (NCA),* Cambridge, MA (pp. 1–10). IEEE.

Misra, S., Deb, P. K., Pathak, N., & Mukherjee, A. (2020). Blockchain-enabled SDN for securing fog-based resource-constrained IoT. *In IEEE INFOCOM 2020-IEEE Conference on Computer Communications Workshops (INFOCOM WKSHPS),* Toronto, ON (pp. 490–495). IEEE.

Nadeem, S., Rizwan, M., Ahmad, F., & Manzoor, J. (2019). Securing cognitive radio vehicular ad hoc network with fog node based distributed blockchain cloud architecture. *International Journal of Advanced Computer Science and Applications,* 10(1), 288–295.

Paul, R., Baidya, P., Sau, S., Maity, K., Maity, S., & Mandal, S. B. (2018). IoT based secure smart city architecture using blockchain. *In 2018 2nd International Conference on Data Science and Business Analytics (ICDSBA)*, Changsha (pp. 215–220). IEEE.

Pedrosa, A. R., & Pau, G. (2018). ChargeItUp: On blockchain-based technologies for autonomous vehicles. *In Proceedings of the 1st Workshop on Cryptocurrencies and Blockchains for Distributed Systems*, Munich (pp. 87–92).

Shrestha, R., & Kim, S. (2019). Integration of IoT with blockchain and homomorphic encryption: Challenging issues and opportunities. *Advances in Computers*, 115, 293–331.

Sookhak, M., Tang, H., He, Y., & Yu, F. R. (2019). Security and privacy of smart cities: A survey, research issues and challenges. *IEEE Communications Surveys & Tutorials*, 21(2), 1718–1743.

Tanwar, S., Parekh, K., & Evans, R. (2020). Blockchain-based electronic healthcare record system for healthcare 4.0 applications. *Journal of Information Security and Applications*, 50, 102407.

Uddin, A.M., Stranieri, A., Gondal, I., & Balasubramanian, V. (2020). Dynamically recommending repositories for health data: A machine learning model. *In Proceedings of the Australasian Computer Science Week Multiconference* (pp. 1–10).

Uddin, M. A., Stranieri, A., Gondal, I., & Balasubramanian, V. (2018). Continuous patient monitoring with a patient centric agent: A block architecture. *IEEE Access*, 6, 32700–32726.

Xie, L., Ding, Y., Yang, H., & Wang, X. (2019). Blockchain-based secure and trustworthy Internet of Things in SDN-enabled 5G-VANETs. *IEEE Access*, 7, 56656–56666.

Yu, Y., Li, Y., Tian, J., & Liu, J. (2018). Blockchain-based solutions to security and privacy issues in the internet of things. *IEEE Wireless Communications*, 25(6), 12–18.

Zhang, P., Liu, F., Kumar, N., & Aujla, G. S. (2020). Information classification strategy for blockchain-based secure SDN in IoT scenario. *In IEEE INFOCOM 2020-IEEE Conference on Computer Communications Workshops (INFOCOM WKSHPS)*, Toronto, ON (pp. 1081–1086). IEEE.

9

Applications of Blockchain Technology in Cyber Attacks Prevention

J. E. T. Akinsola and M. A. Adeagbo
First Technical University

S. A. Akinseinde
The Amateur Polymath

F. O. Onipede
First Technical University

A. A. Yusuf
Federal University of Petroleum Resources

CONTENTS

DOI: 10.1201/9781003193425-9

9.1 Introduction

Blockchain was originally known as Bitcoin Cryptocurrency, but it is much more than that according to Satoshi Nakamoto because it provides exchange of any deal and service over a dispersed system using a reliable template; and it is providing new scopes to productivity of system and security by reforming the digital economy of today (Srivastava et al., 2018). A blockchain is a non-stop rising technology in the list of archives, called block (Eswari, 2017). The blocks are secured, linked and typically contain a timestamp, a transaction data and a link to a previous block called the hash pointer (Mathew, 2019). Blockchain technology offers a technique of recording any digital contact or transaction in a way that is auditable, effective, highly resistant to outages, translucent and safe according to David Schatsky, Managing Director at Deloitte U.S. (Piscin et al., 2017). Blockchain technology eradicates the need of depending on a central consultant by providing independent confirmation assurances. It provides a means of deals division in a wide network of unreliable entities that are simulated and sequentially linked digital ledgers in a distributed database. It is a technology to interrupt and change the future of computing in numerous industries with more advanced results in a revolutionized way (Mathew, 2019). Blockchain services are able to apply immutability against management and misuse if there is a nasty insider, and are able to offer improved security of assets for systems that are dispersed amid different entities (Yassine et al., 2020). Figure 9.1 shows the diagrammatic representation of a simplified look of blockchain architecture.

It is virtually applicable in many environments because it is absolute, circulated and exposed. Cryptocurrencies empowered the technology to gain enormous demand, and it is not only relevant in finance but can also be useful in many other domains (Mathew, 2019) like Internet of Things, supply chain management, reputation systems and healthcare.

A blockchain account deals between noble nodes that do not trust each other because it is a dispersed ledger over a public or private network. Based on a compromise mechanism, the information or data which are added to the chain are confirmed, excavated and chopped into blocks (Srivastava et al.,

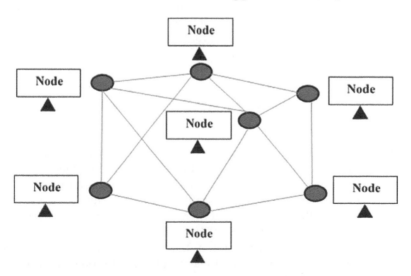

FIGURE 9.1
Diagramatic representation of a simplified look of blockchain architecture.

2018). Compromise mechanisms are designed to allow doubting users in a blockchain network to collaborate and regulate which user submits the next block, Proof of Stake (PoS), Proof of Identity (PoI) or Proof of Authority (PoA), Proof of Elapsed Time (PoET), Proof of Work (PoW) and round-robin. These compromise mechanisms have been used in Bitcoin (Yassine et al., 2020).

One of the general corporate concerns of security organizations is cyber security. The term "cyber security" is frequently used to predict actions to make sure that usage of the Internet safety and security is granted. Also, actions are taking to protect and check against cyber crimes. Cyber crimes are illegal actions which involve the use of computers system or computers as targets and tools (Nwabuike et al., 2020). There is a new opportunity for cyber attackers to engage in adventure because there is high level of addiction on technology and the Internet, and which has led to new business models and income streams for organizations. According to Cyber Risk Lead, there is hopeful revolution in blockchain toward helping enterprises challenge absolute cyber risk tasks while still ensuring data integrity and digital identities (Piscin et al., 2017). Blockchain technology will be very mutual and relevant in qualifying cyber security in the near future (Eswari, 2017).

9.2 Literature Review

9.2.1 Blockchain Evolution

Blockchain was established because of the evolution of virtual currencies and fintech over the last few decades. As an extension of electronic cash system, the widespread use of the Internet favored the emergence of digital currencies at the end of the last century. E-gold, Webmoney, Liberty Reserve and E-cash were established to generate innovative digital currencies (myhsts, n.d.). Blockchain was launched just 10 years ago which affect is possible growth to be a foundation of the universal record-keeping systems. It was created by a person under the pseudonym of Satoshi Nakamoto in which the person was unknown and who was behind the Bitcoin online cash currency (ICAEW, 2021). In 1991, blockchain was initiated and became known to many people by Scott Stornetta and Stuart Haber. Their first work was to develop a method that will ensure that timestamps of documents cannot be tampered with and the method is cryptographically secured using chain of blocks, and Merkle trees was merged into it in order to improve competence, thereby, permitting the assortment of more documents on a single block after upgrading their system (Gwyneth, 2020). Blockchain technology has situated itself since the innovation of Bitcoin, a digital cryptocurrency which is the principal point of notice among various range of practitioners and researchers (Dabbagh et al., 2019).

The first evolution of blockchain is the Bitcoin which does not need any permission from the developer or the user of the system, and it is a public blockchain which was released in 2009 and was legalized by miners using PoW consensus algorithm and this influences the perception of transactions bundled into blocks (Dhaliwal, 2018). Ethereum which was the second evolution of blockchain was announced in 2013 in the form of a decentralized template that operate smart contracts. It explains that blockchain enables developer in storing registries of promises or debts, market creation and moving of resources in accordance without taking any risk of counterparty or middleman. Ethereum is one of the first extensions of blockchain technologies outside of exchange because it is a ledger technology that is used to develop new programs in several companies (Popovski et al., 2018).

9.2.2 Blockchain Platform

The platform has the intension to make all storages that are private such as database to public storage by making all transaction made on the Internet dispersed which could be seen freely under a particular condition (Hanif, 2019). Blockchain platforms enable the blockchain-based applications development which can be either permissionless or permissioned. Some of the platforms that have built blockchain frameworks which permit the user to host and develop applications on the blockchain are Hyperledger, EOS, Ripple, R3 and Ethereum (Takyar, 2020). Table 9.1 shows the comparison between top ten blockchain platforms.

The platform has the intension to convert the client-server architecture into a peer-to-peer (P2P) web that is introduced by Network of Blocks technology as principal storage. There is allocation to the Distributed Hash Table (DHT) on the platform which will account for the storage of data on a wide range of nodes and this will be retrieved with high effectiveness. A model that is multi-level has been introduced by platform architecture, where a piece of node runs on a certain level of the platform under a certain functionality (Hanif, 2019). Figure 9.2 shows network platform blocks in node structure.

9.2.3 Blockchain Architecture

Blockchain consists of blocks that are made consecutively to each other where the whole list of transaction records is found in each block like public ledger that is predictable with a hash block. This is confined in the header of the block and only one parent block is owned by a block. The Ethereum blockchain will store the uncle blocks' (orphan blocks on the Ethereum network) hashes where the genesis block called the first block of a blockchain has no parent block (Lastovetska, 2021). There are three classes where blockchain architecture can fall into. Table 9.2 shows the comparison of the three classes of blockchain architecture.

TABLE 9.1

Comparison of Top Ten Blockchain Platforms (Takyar, 2020)

	R3 Corda	Hyperledger Fabric	Ethereum	Ripple	Open-Chain	EOS	Quorum	Stellar	Hyperledger Iroha	Hyperledger Sawtooth
Consensus Algorithm	Pluggable Framework	Pluggable Framework	Proof of work	Probabilistic Voting	Partionned Consensus	Delegated Proof of stake	Majority Voting	Stella Consensus protocol	Chained-based Byzantine Fault Tolerant	Pluggable Framework
Ledger Type	Permissioned	Permissioned	Permissioned	Permissioned	Permissioned	Permissioned	Permissioned	Both public and private	Permissioned	Permissioned
Industry Focus	Financial Services	Cross-Industry	Cross-Industry	Cross-Industry	Digital Asset Management	Cross-Industry	Cross-Industry	Cross-Industry	Cross-Industry	Cross-Industry
Governance	R3 consortium	Linux foundation	Ethereum developer	Ripple labs	Coin-Prism	EOSIO Core Arbitration Forum (ECAF)	Ethereum Developer and JP Morgan Chase	Stella Development Foundation	Linux Foundation	Linux Foundation
Smart contract	Yes	Yes	Yes	No	Yes	Yes	No	Yes	Yes	Yes

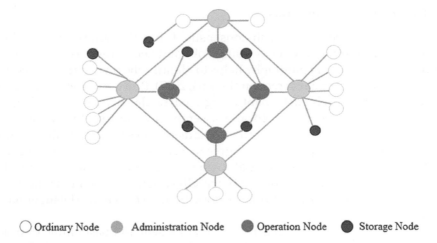

○ Ordinary Node   ● Administration Node   ● Operation Node   ● Storage Node

FIGURE 9.2
Network platform blocks in node structure.

TABLE 9.2

Comparison of the Three Classes of Blockchain Architecture (Lastovetska, 2021)

Property	Consortium blockchain	Public blockchain	Private blockchain
Efficiency	Low	High	High
Immutability	Could be altered	Difficult to alter	Could be altered
Centralized	No	Partial	Yes
Consensus process	Permissioned	Permissioned	Permissioned
Consensus determination	Selected set of nodes	All miners	Single organization
Read permission	Private or restricted	Public	Possibility of restriction or private

9.2.3.1 Blockchain Architecture That Are Public

In the public-based blockchain architecture, the public have access to the system and data or any participant that is willing to participate has access to the system and data, and such systems are Litecoin blockchain, Bitcoin and Ethereum systems (Lastovetska, 2021).

9.2.3.2 Blockchain Architecture That Are Private

The system and the data are controlled only by users that are legal users or from a particular organization that has an invite to participate (Lastovetska, 2021).

9.2.3.3 Blockchain Architecture That Are Consortium

This contains few organizations where preliminary allocated users regulate the setup measures (Lastovetska, 2021).

9.2.4 Blockchain Consensus

Blockchain is used to establish consensus. It has the ability to make historic record prohibitively costly using PoW and keep previous transactions records. Public key cryptography helps to identify the owner and procedure rules help to guarantee correction (Eswari, 2017). Process of consensus' goal is to ensure that each participant using a particular network should agree that the block has been added and assembled according to the network rules and the consensus is the critical characteristics of blockchain because it makes sure that all participants agree to a single "truth" of record (Fuchs, 2019). To adjust data, it is essential to go through a consensus state for the blockchain network. Validation of the modification and automatic checking established over a period of time is done by the consensus (Evangelina & Cano, 2020).

The authors (Reed et al., 2021) stated that present and past online transactions involving digital assets can be proved at any point in time in the future because of its capability of revolutionizing the digital world by enabling a consensus that is distributed. PoW as the consensus model is used by Bitcoin blockchain where the miners are expected to compete against other miners to accept broadcasted and created blocks. Other variations of consensus used by the network for agreement to variations on the ledger are PoS, PoET, PoA and proof of burn (OECD, 2019).

9.3 Cyber Attack Prevention on Blockchain Systems

The following are the solutions provided by blockchain systems to prevent cyber attacks.

9.3.1 Digital Identity

Digital identity (DI) is an entity's online personal data used for granting access to other entities or computing platforms. It is widely used for authentication and data security, though it is not without data privacy concerns and the risk of data breaches through cyber-tracks. The introduction of blockchain to DI poses a formidable solution to the data security risk. Blockchain presents an absolute removal of any form of physical identity, easy accessibility to individuals' identity, data control and management by the entity, and improved data security against cyber attacks and fake identity. A notable use case of DI is Self-Sovereign Identity where entities grant and manage data access to entities (CONSENSYS, 2021).

9.3.2 Decentralized Identifier (DID) Using Cryptography for Data Security

DID is a form of DI that grants an individual or an entity a global unique identifier to be used as an identity to other entities like online platforms, organizations and institutions (Reed et al., 2021).

9.3.3 Phishing Attack

The advancement of the Internet technologies and communication networks has brought about different communication channels for effective dissemination of information among various users. Email is one of the communication channels and indispensable tools employed by individuals and organizations for their efficient operations and correspondence (Cloudphish, 2020). However, the menace of hackers and attacks directed at the system and network (Andryukhin, 2019) especially email is increasingly high. As the attackers create new attacking strategies and enhance the old ones, email is becoming the main entry point into both business and personal affairs for cyber criminal activities. One of the common ways cyber criminals try to intrude the privacy of user credentials and steal information is "phishing" through the use of fake websites or emails (Andryukhin, 2019).

Though there are phishing attacks by social networks, phishing attacks by emails still remain the major risk in any organization because of its indispensable nature (Baillon et al., 2019). A report says that 98% of social incidents and 93% of breaches are through pretexting and phishing with the common vector continuously happens to be email (Vidwans, 2021). Phishing is regarded as an attempt fraudulently carried out by cyber criminals by giving oneself a different appearance or present a malicious electronic communication platform as a trustworthy entity in order to get sensitive information (Ramzan, 2010). Phishing deceptively attempts to acquire financial information such as details of credit card, PIN number, username and password through the use of electronic communication (Baillon et al., 2019).

The three components of phishing attack are vector, medium and approaches (Apandi et al., 2020) otherwise known as schemes (Andryukhin, 2019). Voice, Internet and short messages are the three media cyber criminals used to carry out phishing attack. Among the three media, Internet is regarded the most common medium used by phishers because of its great avenue to launch an attack and easily get at the victims (Chiew et al., 2018; Apandi et al., 2020). The place where the phishing attacks can be launched via the Internet is regarded as vectors which are social networks, email and website (Chiew et al., 2018).

9.3.3.1 Types of Phishing

Conventionally, the approaches attackers use in phishing can be categorized into two types which are technical schemes and social engineering schemes (Andryukhin, 2019).

i. **Social Engineering Schemes**: Social engineering schemes are stealing of sensitive information through a deceptive approach and subsequently use the information to carry out attack on the victim (Andryukhin, 2018). It capitalizes on the emotion of fear of victim to carry out its exploitation, thereby making victim release a sensitive information for phisher to launch an attack (Chiew et al., 2018). At times, users are lured to carry out some actions such as clicking of a link and opening or downloading of malicious documents. This was corroborated in a report on Verizon research that phishing links are clicked by 4% of users (Vidwans, 2021). This enables malware to easily install programs that are malicious into the system and allows phishers to gain entry into the victim computer to carry out a phishing attack (Thakur & Kaur, 2016). Some of the various examples of social engineering schemes are clones phishing, pyramids Ponzi, bloating, fake Initial Coin Offering (ICO), aimed phishing, social networking phishing and fake cryptocurrency wallets.

ii. **Technical Schemes**: Attackers use technical schemes through the deficiencies and vulnerabilities of the infrastructure and software (Andryukhin, 2019). As the software and infrastructure are being improved to detect and prevent phishing attacks, phishers also modify their attack approach by exploiting the vulnerabilities and deficiencies of the infrastructure and software. This attacking strategy is labor-intensive in nature with less visibility, so higher goal percentage is achieved with it than the social engineering. Various examples of technical schemes are session hijacking, malware, Domain Name System (DNS)-based phishing and key loggers.

9.3.4 Mitigating Phishing Attack

In identifying the current trend of phishing, attackers try all possible means to evade the existing defenses which have attracted the attention of researcher to the domain of interest called phishing (Nmachi & Win, 2021). The detection and prevention of phishing is now regarded as a continuous task, and approaches are continuously evolving and growing in quality of being sophisticated. Current approaches proffered by researchers to mitigate attacks through phishing are user classification-based, stylometric analysis, rule-based and user education (Nmachi & Win, 2021). The approaches though have some limitations have metamorphosed into different solutions. The approaches used in the prevention of phishing attacks and their corresponding strengths and limitations are shown in Table 9.3.

TABLE 9.3

Phishing Prevention Approaches and Their Limitations (Nmachi & Win, 2021)

Approaches	Strengths	Limitations
Stylometric analysis	• Reveals identity • Useful for spear phishing and whaling phishing attacks detection	• Change in writing could cause misclassification • Small email size affects accuracy
Rule-based approach	• Performs well on known set rules • Easy to manage	• It has high false alarm rate • Difficulty in updating rules
Classification-based approach	• Can effectively detect phishing emails • Can catch newly create phishing URLs	• It requires a large amount of data and high computational power
User education	• Improves users' ability to detect phishing emails • It educates novice users about phishing attacks	• Lack of knowledge retention • It attracts expense

9.3.4.1 Existing Phishing Solutions and Their Vulnerabilities

Given below are the existing phishing solutions and their vulnerabilities:

i. **Domain-Based Message Authentication Reporting and Conformance (DMARC)**: DMARC is an email policy, authentication and reporting protocol that was built on protocols like Sender Policy Framework (SPF) and Domain Keys Identified Mail (DKIM). DMARC adds author's link to the protocol to monitor, improve and protect the domain against fraudulently sent email (DMARC, 2021). However, with the wide acceptance of the DMARC, the hackers have been increasingly thwarting this protocol. The report has it that, despite the adoption of DMARC, as of 2017, half of the email users still have 16 malicious emails in their inboxes per month. Furthermore, proper configuration of DMARC ensures the entry of email from specific domain can result into wrong rejection of email (Cloudphish, 2020).

ii. **Artificial Intelligence (AI) and Machine Learning (ML)**: AI and ML are employed to prevent the phishing attacks using different ML algorithms (Das et al., 2020). AI is a disruptive technology that is creating a new path for information technology (IT) explorers with the usage of ML (Akinsola et al., 2020). However, it is difficult to differentiate fake email from real email as human writes a phishing email directed at another human, thereby making ML and AI approaches struggle to learn and detect fake emails. ML is developed for helping computer understand the past or present and anticipate or foretell what will occur in future for unidentified situations (Akinsola et al., 2019). Also, ML and AI paradigms are only effective when detecting anomalies in email but fall short in keeping with the pace as per social engineering. An Intelligent User Interface (IUI) anchored on ML paradigms is essential for intelligent phishing solutions (Akinsola et al., 2021).

iii. **Email Filters:** In the email filter, organization considers their requirement for the filtering solution using the combination of various techniques to ensure maximum effectiveness. Among the techniques used are whitelisting, anti-virus, blacklisting, content analysis and gray-listing (Caner Taçoğlu, 2021). However, the filtering system seems to be too aggressive in the blocking of emails thereby causing a communication risk and preventing user from carrying out their job. Also, this phishing solution can only prevent emails that contain malware or Uniform Resource Locators (URLs) that are harmful from getting into the inbox of a user (Cloudphish, 2020).

iv. **User Training:** This is otherwise known as user education (Nmachi & Win, 2021). Security of email using human as agent of prevention is a problem as humans cannot adequately cope with diverse methods of attacks and still require a human to bring forth a solution. Based on the foregoing, some organizations mostly rely on the employees, then train and equip them with the required knowledge to detect phishing attempts. However, this solution is still unable to prevent the phishing attacks as phishing emails that are sophisticated are still being unidentified globally by 97% of people. Therefore, the training of users' needs to be supplemented with improved and appropriate technologies to boost the awareness of employees.

9.3.5 Blockchain Immutability for Phishing Attack Prevention

At present, there is no clear evidence of a tool that can 100% protect users against phishing attack (Sayeed & Marco-Gisbert, 2020) as the lack of means to define the validity of email is the most common challenge in the prevention of cyber phishing attacks. Blockchain with its immutability features can largely and reliably provide the solutions to the lack of means in identifying the validity of email. As evident in the Cloudphish (2020), blockchain technology was used to provide email defense by adding external layer to the security infrastructure of an organization.

Blockchain with its immutable features has a Distributed Ledger Technology (DLT) which was constructed based on P2P network that subsequently creates real-time data that is explicit to its members (Mearian, 2017). It is a digital ledger that is tamper resistant and globally store data without a central repository (Yaga et al., 2018). It has a node that maintains the ledger in the network, and the changes in the ledger must be approved by the nodes before it can be added to the ledger. The immutability nature of blockchain makes transaction in corporate entity or government entity impossible to falsify, manipulate or replace information warehoused on the network (Caner Taçoğlu, 2021). Blockchain technology uses smart contract that outlines the rules and regulations as well as the terms that bind on the contract (Cloudphish, 2020). Blockchain Email Authentication is shown in Figure 9.3.

Blockchain email authentication is central to the security that determines the validity of email. The rules that verify and qualify email as valid enables the prevention of cousin emails, network infiltration, spoofing, account hacking and other unknown types of phishing. As presented in Figure 9.1, email coming from outside of a trusted network or domain will be identified and labeled as unverified through the use of Trusted Sender Authentication Network (TSAN). It uses blockchain techniques to verify that incoming

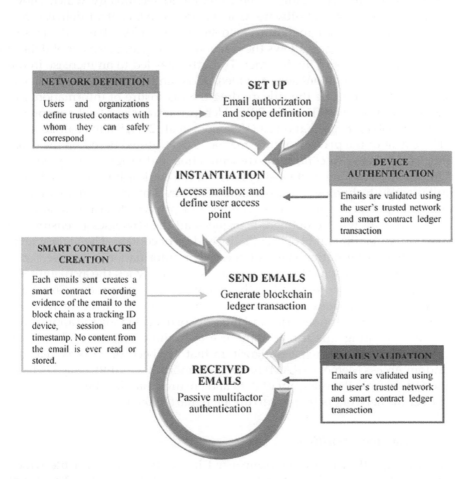

FIGURE 9.3
Blockchain email authentication.

email is originated from a valid device through the use of smart contract in blockchain that is authenticated by the receiver.

9.3.6 Data Security and Data Integrity Attack Prevention Using Hash Function

There are various approaches for data security and integrity attack prevention which are exemplified thus.

9.3.6.1 Data Security and Integrity

The advancement of technologies and emergence of the digital age has brought about a multiplier effect on the number of Internet users across the world. Among the emerging is the cloud-based technology which snowballed into the creation Software as a Service (SaaS), content delivery network, hosting, big data analytics, cloud storage and cloud-based databases. Using these technologies allows the creation, sharing and transfer of data or content among users of the Internet. Thus, this has led to an increase in the number of data or content such as video, photo and messages. As privacy is the ability to determine the rule guiding the disclosure of information and how it should be used (Law360, 2019). Hence, the transfer of data or content must be reliable, secured and fast (Villamora et al., 2019; Salau et al., 2021).

The extent of the privacy of information is determined based on the capability of the owner to control the dissemination and usage of every portion of information as long as it exists. However, the endless business needs and limitations in technologies mostly make high level of privacy impossible to achieve in practical applications. The level of privacy determines how the security of data will be defined. It includes all the strategies to ensure the integrity, availability and confidentiality of data or content. The highest level of security of information is guaranteed when information is accessed only by the authorized user. Also, its integrity is assured when the availability of information to be used is based on the definition of the owner (Law360, 2019).

Furthermore, the possibility of transferring information over the network poses a big threat to the security and integrity of data. Among the approaches used in mitigating this threat is the use of hashing algorithm. The hashing algorithm is a high-speed algorithm that its speed of work is near the Random Access Memory (RAM) limits (Villamora et al., 2019). Usage of hash function controls the integrity of data and ensures that data being transmitted over the network is free of corruption (Duarte, 2008).

9.3.6.2 Hashing Algorithms

Hashing algorithm is an algorithm used in the fixed-size of a file which comprises blocks of information to compute the bit string value. The use of hashing converts the data into shorter length of fixed value that denotes the originally used string. The hash value is a string of characters mostly in a hexadecimal format which is regarded as the distilled summary of the entire content of a file (Chung, 2021). There are many types of hashing algorithms just like an encryption algorithm (Rountree, 2011). Among the most commonly used hashing algorithm are Message Digest algorithm such as MD5, Secure Hash Algorithm (SHA), standards such as SHA-1 and SHA-2, RACE Integrity Primitives Evaluation Message Digest (RIPEMD) and Whirlpool. Others are LAN Manager hashing algorithm (LANMAN), NTLAN Manager algorithm (NTLM) and Cyclic Redundancy Check (CRC) such as CRC32. The resulting output of hash function is referred to as hash value or message digest (Tutorials Point, 2020) as illustrated in Figure 9.4.

Various researchers have employed hash function in different ways to ensure the security and integrity of data. The hash function enables privacy and protection of data which subsequently ensures a reduction of data usage. Ali and Farhan (2020) proposed an enhanced version of Message Digest 5 (MD5) that fortified the algorithm against attacks through the introduction of high efficiency and dynamic variable length into the algorithm. The result of enhanced MD5 shows a high demonstration of resistance against different forms of hacking such as rainbow table attacks, brute force attacks and Christmas attacks.

The detection of any alteration can easily be proven using the hashing algorithm as shown in avalanche affect in the experiment and verification process demonstrated in Almazrooie et al. (2020). This research work employs SHA256 and RIPEMD160, together with a single compression technique that employs two bytes of characters set of Arabic in Unicode UTF-8 and manipulates the data at run time. In the same vein, the authors proposed a framework that can be used in cloud-based Smart Medical System using Elliptic Curve Cryptography (ECC). The proposed framework was adjudged to satisfy various attributes of

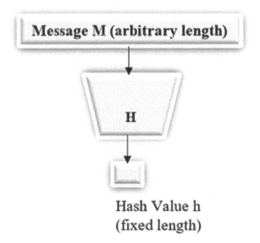

FIGURE 9.4
Hash function.

security such as impersonation attack, doctor anonymity, man in the middle attack, data non-repudiation and data confidentialities to mention but a few.

An unkeyed hash function that utilized the combination arithmetic fixed point and chaotic sponge construction was proposed by Teh et al. (2020) to ensure efficiency and facilitation of upcoming cryptanalysis in the verification of security. The strategies provide a better alternative to floating-point operation which is inefficient and not rely on well-designed paradigm. A verifiable security justification was achieved using sponge construction and the implementation of chaotic map operation through the use of bitwise operation was achieved using arithmetic fixed point.

9.3.7 Properties of a Good Hashing Algorithm

The properties of good hashing algorithm are discussed below.

9.3.7.1 Avalanche Effect

Hashing algorithm is considered to be good when the resulting output of the hash value will entirely or significantly change whenever there is a change even in a single byte or bit of data contained in a file. The inability of hash function to exhibit this property makes the hash value regarded as having poor randomization and has a high tendency of being hacked by the cyber criminals.

9.3.7.2 Unidirectional Process/Pre-Image Resistance

The hash algorithm transforms the content in a file to hash value mostly composed of several string of characters. Such algorithm is considered a good algorithm when the resulting hash output can never be transformed backward to the original data.

9.3.7.3 Complexity/Collision Resistance

An algorithm is considered a good one if the hash algorithm is capable of producing different outputs of hash value when different value is supplied as input. Otherwise, the algorithm will be called hash collision algorithm. Hence, a good algorithm must have a very low tendency of collision.

9.3.8 Benefits of Hash Function

The benefits of hash function are discussed below.

9.3.8.1 Quality of File Comparison

One of the major advantages of hashing is its ability to compare quality of two files without the opening of the file documents for comparison word for word. This is achieved by observing the difference in the computed value of the hash algorithm of the two different files.

9.3.8.2 Verifying the Integrity of File

Hashing algorithm helps in verifying the integrity of file after being transferred over the network or from one location to another especially in a file backup program such as Sync-Back. It ensures non-corruption of transferred file as user can immediately and easily compare their hash value whether they are identical or not.

9.3.9 Application of Hash Function

The various areas in which hash function is applied are discussed thus.

9.3.9.1 Password Storage

Hash function enables the protection of stored password by storing the corresponding hash value during the login of password as against storing it in a clear way. Therefore, user can only see the hash value of password which can neither be used in login process nor retrieve the original value before hashing (Thoma, 2021).

9.3.9.2 Data Integrity

The hash function is the most commonly used application to ascertain the correctness of data by generating the checksum on file that contain the information. It enables user detect every single alteration made, even if it is a byte, on the original file. The process of data integrity occurs at the backend; hence user cannot observe the process. Example of the process of data integrity can be found in messaging application and email system (Sharma, 2020).

9.3.9.3 Proof of Work in Blockchain

Hash function is deployed on blockchain as blockchain utilizes a concept known as PoW. It is imperative that blockchain and Bitcoin technologies must be able to ascertain that users have placed a particular number of computation resources into the process. Therefore, blockchain technologies employ cryptographic hash function for the PoW and request for input generating such hash value that must be in a particular pattern (Thoma, 2021).

9.3.9.4 Digital Signature System

A digital signature is a method that binds an entity or person just like a signatory to the digital information or message. It ensures binding of individual to the data solely as being sent which stored utilizing hash function such that it cannot be stolen or misused and can be subsequently verified at the receiver end (Sharma, 2020).

9.4 Digital Signature

Digital signature over the years has become secured through the introduction of information to the key, utilizing various forms of cryptography and a cutting-edge implementation of signature systems (Heather, 2016). Digital signatures are deployed in the verification of messages or file integrity (Fang et al., 2020). It cryptographically ties users to certain messages just as particular document is linked to a person via signatures. The use of number theory almost makes the forgery of document fortified with digital signature impossible. The number theory uses cryptographic public key in which private key and public key are owned by users to guarantee security functionality (TAM, 2021). The secret information that enables the proves of public key ownership is regarded as private key while the key that proves owner's identity is regarded as public key. These private and public keys form a pair and are both owned by a user. The non-value transfer capability of digital signature and carriage of information only on the Internet make it applicable and more secured when compared with traditional application (Fang et al., 2020).

The method of digital signature provides systems with non-repudiation, integrity and validation of access to data contents electronically sent across the networks. The mathematical algorithm used in digital signature uses information in message content and stored information in the key to create signature otherwise called hash. Digital signature is employed using mathematical algorithm to enable the control of organizational security of system such as emails and provide authentication and protection of data against forgery. The algorithm used in digital signature is generally made up of three phases which are key generation, signature generation and verification of algorithm (Pallipamu et al., 2014). The techniques employed in digital signature can be broadly categorized into two which are Direct digital signature and Arbitrated digital signature (Roy & Karforma, 2012). The first technique allows the exchange of messages only between the sender of message and the receiver while the latter allows the exchange of messages through trusted third party only between the sender of message and the receiver. Digital signatures had been applied in various domains to ensure authentication of access and data integrity. Among the domains are fault tolerance, contract signing protocol, object-oriented software engineering, web-based assessment system, key agreement protocol, chip level programming and identity-based authentication.

9.4.1 Applications of Digital Signature

Digital signature is applicable in various domains, which are discussed below.

9.4.1.1 Key Agreement Protocol

The key agreement protocol allows two or more entities confidentially exchange messages through the use of shared encrypted key (Doubleoctopus, 2020). The encrypted key in the key exchange protocol utilizes various cryptographic techniques to achieve this goal. Therefore, for two entities to exchange information confidentially, the secret key that encrypt and decrypt messages must have been firstly exchanged before communication can occur (Just, 2011). These approaches are also being deployed using the Elliptic Curve Version to enhance Digital Signature in the area of E-voting, E-governance, E-learning and E-shopping to mention but a few (Roy & Karforma, 2012).

9.4.1.2 Contract Signing Protocol

Digital signature techniques are employed to overcome problem of suspicion between two parties in carrying out mutual contract relationship with minimal risk electronically. This technique is deployed in electronics transaction (Ferrer-Gomila & Hinarejos, 2020) such as E-commerce, online payment of examination fees and tax payment systems to mitigate mutual distrust between parties. It initiates quality of service agreement called Service Level Agreement (SLA) that maintains and specifies penalty provision in case of any breach of contract.

9.4.1.3 Fault Tolerance

Fault tolerance describes the techniques to ensure possible provision of services that bring about reliable software even in the incidence of faults (Imperva, 2021). The state of fault tolerance is achievable through processing of error or fault treatment that snowballed into the creation of sophisticated software that are less susceptible to interruption of hardware or software during the implementation. Error compensation or error recovery are employed in the software using error processing with a view to removing software errors while the prevention of fault activation before the creeping in of errors are achieved using fault treatment.

9.4.1.4 Web-Based Assessment System

This allows provision of new tools for education and thus ensures that the community of researchers have more information and greater flexibility. ECC is sometimes embedded to enable easy spread of education and study materials such as video or audio material. The purpose of ECC is to ensure the originality and integrity of study material being sent to students from the real sender.

9.4.1.5 Identity-Based

This is employed in the identity card such as smart card application used in banking, insurance, education and place of work. It uses digital signature algorithms that are easily implemented in the system that uses asymmetric-based identity key pair as against the traditional private/public key pair. The asymmetric-based identity key pair is induced into the protocol to ensure the establishment of key authentication and for proper identification.

The digital signature has been proven as one of the approaches to ensure authentication of access and integrity of data with its capability of cost efficiency, flexibility, non-repudiation, time efficiency and imposition of industrial standards. However, despite the humongous security advantages embedded in digital signature and the future opportunity in retrospect, there exist some security issues which can be surmounted using blockchain technology.

9.4.2 Application of Blockchain on Digital Signature

Digital signature that utilizes Verifiable Encrypted Signature requires an adjudicator. There exist various security challenges regarding the use of adjudicator which is a centralized trusted third party (TTP). Examples of such are interruption of services, connivance with others and disclosure of contract contents. However, these problems of centralized trusted third party (TTP) can be addressed using smart contract approach in blockchain. Smart contract in blockchain technology provides impartial contract signing techniques between parties to allow fair signing of contract among participants in the blockchain. Blockchain technology discourages centralization and encourages decentralization with autonomy, tampering resistant, verifiability and efficiency (Fang et al., 2020). The use of blockchain solution enables publishing of data at suitable cost. Examples of the application of blockchain on digital signature are Ethereum and Bitcoin solutions such as Dash and Litecoin (Ferrer-Gomila & Hinarejos, 2020). Therefore, blockchain are used to enhance digital signature with a view to avoid suspicion in any electronic transaction between entities and discourage the use of centralize TTP.

9.4.3 Blockchain Immutability for Phishing Attack Prevention

Blockchain has no distinct place for someone to initiate a fraud scheme because management and permission is spread across the network. Transactions stored in blockchain cannot be altered because of its immutability. Blockchain technology develops a process called consensus, which the network associate must agree that the transaction is confirmed before a block of transactions attached to the blockchain can be endorsed. Endorsement comes in place because there must be some ways to ensure that outsiders cannot corrupt the voucher which helps in preventing phishing attack. Phishing is a prominent cyber attack that employs impersonation and social engineering to breach user's data. The keys in endorsed network control access and identity management. Most phishing attempts are largely due to human factors, which have been the Achilles Heel of cyber security.

9.5 Best Practices for Mitigating Injection Attack Using Blockchain

A method that is used by attackers to supplement SQL (Structured Query Language) query into input fields which is then processed by the basic SQL database is known as SQL injection (Positive Technologies, 2021). It is an injection that uses code method in which malicious SQL statements are inserted into the entry field for execution in attack data-driven applications (Wikipedia, 2021).

SQL injection is an unforeseen consequence or attack like authorization of mechanisms and circumvention of authentication that occurs when specifically constructed input can provoke an application into mis-constructing a database command allowing modification, deletion, retrieving and addition of attack and records compromising the integrity of a database and the applications it provides (Chen et al., 2018). Availability, integrity, confidentiality and functionality of any web application databases are threatened by Structured Query Language Injection Attack (SQLIA). Dynamic leakage, static leakage, spoofing and linkage leakage are other examples of attack that endanger database security (Abimbola & Zhangfang, 2020).

It would be problematic to point any changes to a user who is fortunate to gain access into the database and alters the corresponding log accesses as a result of access control in most of the centralized database systems and when applications are based on organizational polices which is an example of insider attack and it might be difficult to sense such attack. A suggested method or technology used for sensing such an attack is blockchain. The metadata will help us to validate the database accesses, but it would have a small overhead storage (S. Sharma et al., 2017).

Proxy server is one of the methods introduced to eradicate SQLIA. It is planted between two devices which is used for communication and is used to filter any attempt of SQL injection. Figure 9.5 shows the architectural model of proxy server. The procedures to follow in proxy server are as follows: a list of common commands of SQL should be constructed, proxy server that will alert the administrator of database of possible SQL injection commands should be created, SQL query command structures should be analyzed, making sure that filter developed to work in the proxy server proves its ability to prevent SQL injection, and SQL injection attack on the database is prevented using proxy server. The resultant effects of the attacks are visible with SQLIA (Idowu et al., 2020).

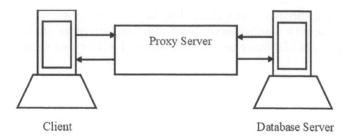

FIGURE 9.5
The architectural model of proxy server.

The disadvantages of this method which leads to the development of another method is that authentic word can be filtered from the variables during the process of filtering, and it cannot be used if the data is encrypted because without decryption it is impossible to view in plain text. Another method for mitigating injection attack using blockchain is by using one of the blockchain platforms called stellar. Stellar is a developer-friendly, decentralized and open-source platform that allows anybody to trade, issue resources and settle expenses. Signature system used is Ed25519 public key. Only thing required from stellar is the keypair. The digital signature structure used is Edwards curve Digital Signature Algorithm (EdDSA) where a variant of Schorr signature based on Twisted Edwards curves is used (Curve25519 and SHA512), which is considered to be quicker without losing security. In order to make it impossible for hackers to get access to the entire process, the public keys are authenticated to private key. A proxy server is planned to be positioned among the database server and the client where the keypair is produced (Abimbola & Zhangfang, 2020).

9.5.1 Innovative Use of Blockchain Technology for Data Security

Innovation refers to as the execution and beginning of important variations in the process, organization or marketing the product of the company in order to cultivate the result of the company. Innovation precedes changes by application of expertise and knowledge which can be established or attained by purchase of technology or advisory services of external collaboration (Evangelina & Cano, 2020). Innovation also refers to innovative and unique ways used by organizations to estimate, execute and articulate their actions and rules in order to attain competitive benefits that are justifiable (Mohamud, 2020). Other areas where blockchain can be applied are shown in Figure 9.6.

9.5.1.1 Cyber Security

Disseminating threat information can be used to combat future cyber attacks using blockchain (Rawat et al., 2020). Blockchain can offer the checking and transparency of communication security, transaction, security of user uniqueness, and the critical infrastructure protection. It makes use of shared

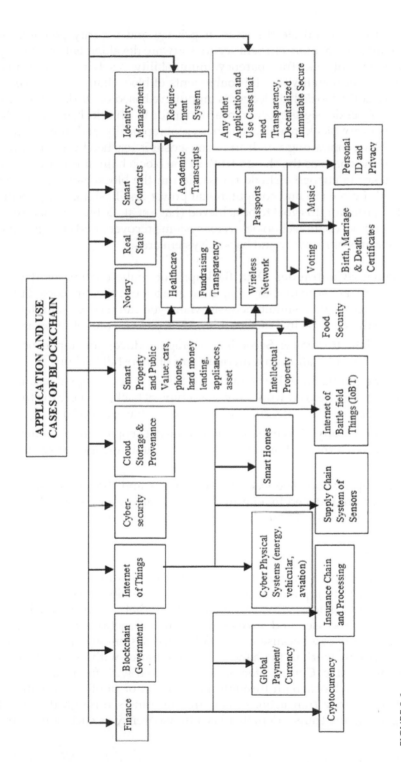

FIGURE 9.6
Application areas of Blockchain.

online services to eradicate the issue of privacy and security potential. It helps in realization of more efficient secure software development and supply chain management risk. Transparency, immutability, resolving trust, transactions and accountability are the components of blockchain in cyber security (Eswari, 2017). Blockchain ledger is immutable based on its data integrity component. A sender signs every transaction made in a block chronologically, miner signs every block in a blockchain cryptographically, and immediately, preceding block containing a hash and a consensus is reached by all the applicants about the chain as the shared truth in the blockchain. It is not possible for an attacker to change a single transaction in the blockchain because of the number of electrical and computational power and hashing components required in order for this goal to be achieved (Lage et al., 2019). Another innovative use of blockchain technology is for systems recovery and storage security. The features of recovery and backup system are:

i. **Automatic/Continuous Data Backup**: Automatic/continuous data backup makes sure that any changes made to your files are copied to the storage location simultaneously where you can recover your data in case there is loss in data, but it lowers objective point of recovery.

ii. **Immediate Recovery**: Secondary storage reduces an application downtime and it allows a temporary run of snapshot backup.

iii. **Incremental Backup**: This reduces the time it takes to copy data and it does not make your work slow. It is a type of backup where the full file is not copied but the changes made to the file are copied.

iv. **Error-Free Copy**: This makes sure that the data copied from a source and stored at the backup server does not contain errors and are the same.

v. **Data Duplication**: The load on the network and the storage space required are reduced by eliminating duplicated data record blocks and data is transferred to the storage backup location.

9.5.1.2 Finance

Blockchain is adopted in most banking organizations for extensive series of purposes including payment dispensation, transaction cross-border and trade settlements (Fuchs, 2019). An intermediary that is conventional such as banks that processes and verifies financial transactions makes use of blockchain technology because it is a centralized system and which makes putting enormous work in the hand of intermediaries without any error in their output possible. Complicated issues associated with financial services are simplified by blockchain through introduction of public ledger that is distributed to verify transactions made by the miners through the use of PoW. Regarding the transaction, there is a transparency as each node in the blockchain network contains a copy of the blockchain that has been updated. Once a block with a confirmed transaction is added to the blockchain, the complete blockchain is unchallengeable as the blocks are organized chronologically (Rawat et al., 2020). It is also applied financially in Bitshares, Medici, Private Securities, Blockstream, Augur, Coinsetter, etc. (Crosby et al., 2015).

9.5.1.3 Government

Blockchain technology is applied in government units and organization to build operative and reliable government operation through transparent and cooperative network (Rawat et al., 2020). Information is published on improved and original influence agreements by the National Research Council using Ethereum smart contracts for more transparency and better public programs functionality (Kramer, 2019). Patents, individual identities, property record and other primary data are made possible due to blockchain immutability, single source of truth and chronological of the data (Fuchs, 2019). In order for the government to address the accountability of its bodies, its affairs must be transparent and in order for the government to do so, it must make a great amount of data open to the public. Application of Blockchain in government is an alternative to this, where the data is open to the public all the time because it is distributed in the public ledger. To make sure the electorates work in the favor of the electors, smart contracts can be used. The electorates only get paid once they meet the demand of the electors via the smart contracts (Rawat et al., 2020).

9.5.1.4 Internet of Things (IoT)

The ideas of smart home are feasible because the way of lives is expected to be transformed using IoT. Grave problems concerning privacy and cyber security is created through the use of this new phenomenon because of its ability of having enormous number of connected devices and making lives comfortable. The important technology that can be used to secure IoT is Blockchain Technology. In order to make sure that information passing through IoT will ensure accountability in addition to security of the participants, it is important to make use of blockchain technology because millions of devices are communicating and linked to each other (Rawat et al., 2020). Blockchain technology facilitates the usage of decentralized IoT platforms, for example, the data trusted in addition to secured data exchange as well as efficient record keeping (Crosby et al., 2015). Blockchain have the ability to manage Machine-to-Machine (M2M) communication by making use of its automation and speed capabilities where the machine will have the ability to communicate or contact each other without any contact or command from the human (Evangelina & Cano, 2020).

9.5.1.5 Cloud Storage and Provenance

Executing a peer-to-peer cloud storage network offers a solution to traditional storage network challenges by providing end-to-end encryption, where users can share and transfer data securely without trusting on a third party for consistency and safety (Crosby et al., 2015). Integrating with blockchain technology can help to get rid of security and data privacy issues in order to facilitate cloud storage growth. Blockchain technology helps to manage data and improves service accessibility and security of data (Murthy et al., 2020).

9.5.1.6 Smart Contracts

A set of rules are stored on the blockchain to speed transactions and must be automatically executed. The contracts are routinely executed when the conditions are satisfied, and because of the use of blockchain, the smart contract confirms that the contract particulars are known by the participants. Before transactions are appended and converted to a new block in the blockchain, the transactions are broadcasted into the blockchain network and they are verified and collected by the miners. The miners are collections of common peers who validate the transactions associated to contract in order to make the smart contracts works (Rawat et al., 2020). All possible uncertainties concerning the implementation of the contract conditions are eliminated and the process of negotiation are made more efficient and easier through the use of smart contracts which takes after the contractual clauses and makes sure the interface is usually clear purposely in logic to decrease the cost of agreement and to confirm that the predetermined processes are protected (Mohamud, 2020). A smart contract guarantees that the network is decentralized because it only works if there is no partiality in the computational power of the network and it solves any possible battle through the use of consensus procedure which is created on PoW (Rawat et al., 2020).

9.5.1.7 Healthcare

Sensitive information like personal health record of the patient needs to be kept with high level of security. Personal records can be kept and determined through the use of blockchain and making sure that only precise individuals have access to the records by using private key provided by blockchain. Research can be conducted using the same procedure where records that are personal are used by Health Insurance Portability and Accountability (HIPAA) laws which are used to ensure data confidentiality (Rawat et al., 2020). Huge sums of administrative and clinical data from medicinal stock chain to entitlements management and patient medical records are achieved by industry of worldwide healthcare. Everything connected to surgical suites to personal fitness trackers are entirely presented to new ecosystem of information to the pool of data collected and mined which is growing exponentially through the introduction of smart medical devices (Fuchs, 2019). It allows information to be shared between stakeholders of different healthcare systems and information are shared from end users to new earners easily (OECD, 2019).

9.5.2 Blockchain Impact

Blockchain has the ability to make transactions over the Internet firmly between two parties without intrusion of any intermediate party (Piscin et al., 2017). Blockchain is used by government in establishment of trust, prevention of fraud, improvement of transparency and speed up the rate of transactions (Eswari, 2017). Blockchain is a technology where noble nodes do not trust each other which allows it to account for atomicity, integrity and auditability of data and makes it achieve reliability in transactions over dispersed independent platforms (Srivastava et al., 2018). It has the ability to verify past and present online transactions involving digital assets in the future and to revolutionize the digital world by allowing a distributed compromise which is done without compromising the secrecy of the digital assets and parties involved (Crosby et al., 2015). Blockchain enables the process of tracking assets and recording transactions in a business network because it is a circulated and collective ledger (Eswari, 2017). Blockchain was applied and executed in smart contract which implements the terms of a contract in an electronic transaction procedure. Smart contract can be implemented automatically by miners because it uses an encryption fragment and it has the ability to transform several fields like IoT and financial services (Zheng et al., 2017). Blockchain discourages cyber criminals from their ventures by making cyber crimes costly to predators through the use of cryptography (Nwabuike et al., 2020).

9.6 Conclusion

Blockchain technology is undoubtedly a veritable tool that is applicable in various industries and has been a great influence on cyber security specifically to mitigating recent hacker attacks around the globe. The applications of blockchain in DI have helped in the prevention of inaccessibility, data insecurity and fraudulent identities with key benefits to organizations, users as well as IoT coordination systems. Cyber attacks can be mitigated with the use of blockchain technology in relation to data security and integrity attack, DI attack, phishing attack as well as injection attack.

References

Abimbola, J. O., & Zhangfang, C. (2020). Prevention of SQL injection attack using blockchain key pair based on stellar. In *European Scientific Journal ESJ, 16*(36), 1–13. doi: 10.19044/esj.2020.v16n36p92.

Akinsola, J. E. T., Adeagbo, M. A., & Awoseyi, A. A. (2019). *Breast Cancer Predictive Analytics Using Supervised Machine Learning Techniques. 8*(6), 3095–3104.

Akinsola, J. E. T., Awodele, O., Idowu, S. A., & Kuyoro, S. O. (2020). SQL injection attacks predictive analytics using supervised machine learning techniques. *International Journal of Computer Applications Technology and Research, 9*(4), 139–149. doi: 10.7753/ijcatr0904.1004.

Akinsola, J. E. T., Akinseinde, S., Kalesanwo, O., Adeagbo, M., Oladapo, K., Awoseyi, A., & Kasali, F. (2021). Application of artificial intelligence in user interfaces design for cyber security threat modeling. In *Intelligence User Interface* (pp. 1–28). IntechOpen, London.

Ali, A. M., & Farhan, A. K. (2020). A novel improvement with an effective expansion to enhance the MD5 hash function for verification of a secure E-Document. *IEEE Access, 8*, 80290–80304. doi: 10.1109/ACCESS.2020.2989050.

Almazrooie, M., Samsudin, A., Gutub, A. A. A., Salleh, M. S., Omar, M. A., & Hassan, S. A. (2020). Integrity verification for digital Holy Quran verses using cryptographic hash function and compression. *Journal of King Saud University - Computer and Information Sciences, 32*(1), 24–34. doi: 10.1016/j.jksuci.2018.02.006.

Andryukhin, A. A. (2018). Methods of protecting decentralized autonomous organizations from crashes and attacks. *Proceedings of the Institute for System Programming of the RAS, 30*(3), 149–164. doi: 10.15514/ispras-2018-30(3)-11.

Andryukhin, A. A. (2019). Phishing attacks and preventions in blockchain based projects. *International Conference on Engineering Technologies and Computer Science*, 15–19. doi: 10.1109/EnT.2019.00008.

Apandi, S. H., Sallim, J., & Sidek, R. M. (2020). Types of anti-phishing solutions for phishing attack. *The 6th International Conference on Software Engineering & Computer Systems*, 1–9. doi: 10.1088/1757-899X/769/1/012072.

Baillon, A., De Bruin, J., Emirmahmutoglu, A., Van De Veer, E., & Van Dijk, B. (2019). Informing, simulating experience, or both: A field experiment on phishing risks. *PLoS One, 14*(12), 1–15. doi: 10.1371/journal.pone.0224216.

Caner Taçoğlu. (2021). *Email Filtering.* Proofpoint.Com.

Chen, G., Xu, B., Lu, M., & Chen, N. (2018). *Exploring Blockchain Technology and Its Potential Applications for Education.* 1–10. doi: 10.1186/s40561-017-0050-x.

Chiew, K. L., Yong, K. S. C., & Tan, C. L. (2018). A survey of phishing attacks: Their types, vectors and technical approaches. *Expert Systems with Applications, 106*, 1–20. doi: 10.1016/j.eswa.2018.03.050.

Chung, C. (2021). *Introduction to Hashing and Its Uses.* 2BrightSparks Pte. Ltd., Singapore.

Cloudphish. (2020). Blockchain prevents phishing attacks other technologies miss. In cloudphish.com.

CONSENSYS. (2021). Blockchain for digital identity: Real world use cases | ConsenSys. CONSENSYS.

Crosby, M., Nachiappan, Pradhan, P., Verma, S., & Vignesh, K. (2015). BlockChain Technology. In Sutardja Center for Entrepreneurship & Technology Technical Report.

Dabbagh, M., Sookhak, M., & Safa, N. S. (2019). The evolution of blockchain : A biblio-metric study the evolution of blockchain : A bibliometric study. *IEEE Access, 7*, 19212–19221. doi: 10.1109/ACCESS.2019.2895646.

Das, M., Saraswathi, S., Panda, R., Mishra, A. K., & Tripathy, A. K. (2020). Exquisite analysis of popular machine learning–based phishing detection tech-niques for cyber systems. *Journal of Applied Security Research*, 1–25. doi: 10.1080/19361610.2020.1816440.

Dhaliwal, J. (2018). The evolution of blockchain.

DMARC. (2021). What is DMARC? Dmarc.Org.

Doubleoctopus. (2020). Key agreement protocol. The Secret Security Wiki.

Duarte, R. P. (2008). Transport protocols for large bandwidth-delay product net-works - TCP extensions and alternative transport protocols. *In Workshop on Electronics, Telecommunications and Computers Engineering* (Issue November 2008). Lisbon, Portugal

Eswari, P. R. L. (2017). Blockchain technology and its significance to cyber security. *International Conference on Public Key Infrastructure and Its Applications*, 1–64. Bangalore, India

Evangelina, J., & Cano, S. (2020). The technological innovation of the blockchain and its impact on the energy sector (La innovación tecnologica del blockchain y su impacto en el sector energético). *XVI*, 157–178.

Fang, W., Chen, W., Zhang, W., Pei, J., Gao, W., & Wang, G. (2020). Digital signature scheme for information non-repudiation in blockchain: a state of the art review. *Eurasip Journal on Wireless Communications and Networking*, (1), 1–15. doi: 10.1186/s13638-020-01665-w.

Ferrer-Gomila, J. L., & Hinarejos, M. F. (2020). A 2020 perspective on "A fair contract signing protocol with blockchain support." *Electronic Commerce Research and Applications, 42*, 1–3. doi: 10.1016/j.elerap.2020.100981.

Fuchs, P. (2019). Blockchain. *MERCER*, 1–25.

Gwyneth, I. (2020). Blockchain technology history: Ultimate guide. In *101 Blockchains*.

Hanif, M. (2019). Blocks' Network: Redesign architecture based on blockchain technology [Embry-Riddle Aeronautical]. *In Proceedings - IEEE International Conference on Industrial Internet Cloud, ICII 2019*. doi: 10.1109/ICII.2019.00017.

Heather, W. (2016). How digital signatures and blockchains can work together. Cryptomathic.Com.

ICAEW. (2021). History of blockchain _ technology _ ICAEW.

Idowu, S. A., Awodele, O., Kuyoro, S. O., & Akinsola, J. E. T. (2020). Taxonomy and char-acterization of structured query language injection attacks for predictive analyt-ics. *International Journal of Software & Hardware Research in Engineering (IJournals), 8*(3), 15–25. https://www.academia.edu/download/65420840/IJSHRE_8306_Taxonomy_and_Characterization_of_SQLIA_Idowu_Sunday_A.1_Awodele_Oludele2_Kuyoro_Shade_O.3.pdf.

Imperva. (2021). What is fault tolerance. Imperva.Com. doi: 10.1145/163640.163645.

Just, M. (2011). Key Agreement. In *Encyclopedia of Cryptography and Security* (pp. 319–344). doi: 10.1007/0-387-23483-7_218.

Kramer, M. (2019). An overview of blockchain technology based on a study of public awareness. *Global Journal of Business Research 13*(1), 83–91.

Lage, O., Diego de, S., Urkizu, B., & Gomez, E. (2019). *Blockchain for Cybersecurity and Privacy_ Architectures, Challenges, and Applications*.

Lastovetska, A. (2021). Blockchain architecture explained how it works & how to build. In *MLSDev.*

Law360. (2019). The privacy questions raised by blockchain. Bradley Arant Boult Cummings LLP.

Mathew, A. R. (2019). Cyber security through blockchain technology. *International Journal of Engineering and Advanced Technology, 9*(1), 3821–3824. doi: 10.35940/ijeat.A9836.109119.

Mearian, L. (2017). What is blockchain and how can it help business?

Mohamud, A. A. (2020). Blockchain technology as an innovation strategy for competitive advantage by kenya redcross society. University of Nairobi.

Murthy, B., Shri, L., Kadry, S., & Lim, S. (2020). Blockchain based cloud computing : Architecture and research challenges. *Researchgate.* doi: 10.1109/ACCESS.2020.3036812.

myhsts. (n.d.). *History and Evolution of Blockchain Technology from Bitcoin | Blockchain Developer Guide.* Myhsts. Retrieved July 8, 2021, from http://myhsts.org/tutorial-history-and-evolution-of-blockchain-technology-from-bitcoin.php.

Nmachi, W. P., & Win, T. (2021). Phishing mitigation techniques: A literature survey. *International Journal of Network Security & Its Applications, 13*(2), 63–72. doi: 10.5121/ijnsa.2021.13205.

Nwabuike, C. C., Onodugo, V. A., Arachie, A., & Nkwunonwo, U. C. (2020). Blockchain technology for cyber security : Performance implications on emerging markets multinational corporations, overview of nigerian. August.

OECD. (2019). OECD Blockchain Policy Forum OECD Blockchain Primer Policy. OECD.

Pallipamu, V. R., Reddy, T., & Varma, S. (2014). A survey on digital signatures. *International Journal of Advanced Research in Computer and Communication Engineering, 3*(6), 7243–7246.

Piscin, E., David, D., & Kehoe, L. (2017). Blockchain & cyber security. Let's discuss for more information please contact . *Eloitte,* 1–14.

Popovski, B. L., Soussou, G., & Webb, P. B. (2018). A brief history of blockchain.

Positive Technologies. (2021). How to prevent SQL injection attacks. Positive Technologies.

Ramzan, Z. (2010). Phishing attacks and countermeasures. In P. Stavroulakis & M. Stamp (Eds.), *Handbook of Information and Communication Security* (pp. 433–434). Springer. doi: DOI: 10.1007/978-3-642-04117-4_23.

Rawat, D. B., Chaudhary, V., & Doku, R. (2020). Blockchain technology: Emerging applications and use cases for secure and trustworthy smart systems. *Journal of Cybersecurity and Privacy, 1*(1), 1–15. doi: 10.3390/jcp1010002.

Reed, D., Sporny, M., Sabadello, M., Dave, L., & Christopher, A. (2021). Decentralized Identifiers (DIDs) v1.0. W3C.

Rountree, D. (2011). Cryptography. In *Security for Microsoft Windows System Administrators* (pp. 29–69). Elsevier Inc. doi: 10.1016/B978-1-59749-594-3.00002-8.

Roy, A., & Karforma, S. (2012). A survey on summarizers and its applications. *International Journal of Current Engineering and Technology, 3*(1), 45–69.

Salau A. O., Marriwala N., & Athaee M. (2021). Data security in wireless sensor networks: Attacks and countermeasures. In *Lecture Notes in Networks and Systems* (vol. 140, pp. 173–186). Springer, Singapore. doi: 10.1007/978-981-15-7130-5_13.

Sayeed, S., & Marco-Gisbert, H. (2020). Proof of Adjourn (PoAj): A novel approach to mitigate blockchain attacks. *Applied Sciences (Switzerland)*, *10*, 1–23. doi: 10.3390/APP10186607.

Sharma, M. (2020). Applications of Hash Function in Cryptography. Includehelp. Com.

Sharma, S., Gupta, R., Srivastava, S. S., & Shukla, S. K. (2017). Detecting insider attacks on databases using blockchains. *Workshop on Blockchain Technologies and Its Applications*. ISRDC, IIT Bombay, Bombay, India

Srivastava, A., Bhattacharya, P., Singh, A., & Mathur, A. (2018). A systematic review on evolution of blockchain generations. *International Journal of Information Technology and Electrical Engineering*, *7*(6), 1–8.

Takyar, A. (2020). Top blockchain platforms of 2020 for blockchain application. In World Wide Web. LeewayHertz.

Tam, A. (2021). Digital signatures in blockchains: The present and future. Coinbase Crypto Services.

Teh, J. Sen, Alawida, M., & Ho, J. J. (2020). Unkeyed hash function based on chaotic sponge construction and fixed-point arithmetic. *Nonlinear Dynamics*, *100*(1), 713–729. doi: 10.1007/s11071-020-05504-x.

Thakur, H., & Kaur, S. (2016). A survey paper on phishing detection. *International Journal of Advanced Research in Computer Science*, *7*(4), 64–68. doi: 10.26483/ijarcs.v7i4.2706.

Thoma, M. (2021). The 3 applications of Hash functions. Gitconnected.Com.

Tutorials Point. (2020). Cryptography Hash functions. Tutorialspoint.Com.

Vidwans, R. (2021). 5 biggest data breaches of all time from phishing. Clearedin.

Villamora, V. J. A., Lim, M. S., & Sebial, A. J. (2019). Modern file transfer protocol using lossless compression, lattice-based encryption, and a data integrity hashing function. *IOP Conference Series: Materials Science and Engineering*, *482*(1), 0–6. doi: 10.1088/1757-899X/482/1/012050.

wikipedia. (2021). SQL injection. In Article.

Yaga, D., Mell, P., Roby, N., & Scarfone, K. (2018). Blockchain technology overview. In NIST Interagency/Internal Report. doi: 10.6028/NIST.IR.8202.

Yassine, M., Alazab, M., & Romdhani, I. (2020). Blockchain for cybersecurity and privacy. In *Blockchain for Cybersecurity and Privacy*. doi: 10.1201/9780429324932.

Zheng, Z., Xie, S., Dai, H., Chen, X., & Wang, H. (2017). *An Overview of Blockchain Technology : Architecture, Consensus, and Future Trends*. 557–564. doi: 10.1109/BigDataCongress.2017.85.

10

A Lightweight Digital Voting Platform Using Blockchain Technology

Nithin Kamineni, Veera Nitish Mattaparthi, A. Mona Reddy,
T. Mahalakshmi, Vamsi Pachamatla, Kuldeep Chaurasia,
and Tanmay Bhowmik
Bennett University

CONTENTS

10.1 Introduction

As Blockchain technology emerged, the main idea of decentralization has been developed. The aim of this research is to build an e-voting platform as a more reliable and secured application of Blockchain technology. The business sector and service sectors of society, such as the insurance and manufacturing industries convey confidential data with the help of a trusted third party. This process is very complicated and hence they are facing several issues. Since the features of decentralization are mostly found and used in Blockchain technology the researchers came up with the idea of integrating the original design of the e-voting system with Blockchain. This increases data security, provides voter anonymity and lowers election costs, and thus preserves the fairness and integrity of the vote. Through an encryption scheme of Paillier public-key cryptosystem privacy and data protection can be gained.

The transparency of Blockchain helps in verifying the vote results by the voters without the need of a trusted third party. Crypto currency bitcoin forms the fundamental of Blockchain technology. It is a public ledger and consists of a distributed database, where records are stored in the form of transactions and a block is a set of such transactions. The architecture of Blockchain technology is designed in such a way that it has various locations in its distributed database so there is no issue of a single point failure. It has shared control over who can add transactions to the database. Whenever a new block is added to the Blockchain, it refers to the address of the preceding version of the database; hence this generates an unchangeable model and also prohibits the unauthorized alteration of the previous entries. Before the appended latest block is converted into a permanent member of the ledger, a huge proportion of the network nodes have to reach coherence. Hence compared to any other database, these innovative cryptographic characteristics offer enhanced security making the Blockchain technology ideal for e-voting. Manuscript is organized as follows.

In Section 10.2, we have discussed the Related Work and researches done in the past based on e-voting schemes via decentralized network. Section 10.3 depicts the various data resources of the project, whereas Section 10.4 throws some light on design architecture and methodology. The outcomes of our system are shown in Section 10.5. Section 10.6 gives the details of the performance evaluation. Finally, the conclusion of our project and our acknowledgments toward people and authorities, several referred papers, articles, and websites are available in Section 10.7.

10.2 Related Work

In the past, there have been many attempts in making e-voting systems, but still they have been complex and less secure. Hence, this can be perfectly improved by using Blockchain technology. As per our research in previous works, we have found that Shamir [1] first suggested the idea of "Hidden sharing." It provides for efficient protection from server-side assaults. In 1981 an encrypted voting stream was used by Chaum through a cryptographic platform to encrypt the ballot. DeMillo et al. [2] introduced a framework requiring all electors to participate and the ballot of every voter would be encrypted. As cryptography advanced, a lot of schemes with unique properties were introduced. Later a realistic hidden e-voting system was suggested for elections held in big scale that would guarantee the privacy of citizens by Fujioka et al. [3].

The concept of blind signature was used to mask the vote of the voter and was then delivered to the administrator. After this many software related to digital voting were introduced in the market, and industries like Evox and Sensux used them. But the flaw in this system is that all citizens must vote and it is a compulsory requirement. If anyone dissents or doesn't vote the entire election outcomes are changed or affected. The programmer or the administrator cannot figure out the person influencing the outcome. This scheme was also modified in Ref. [4] by using the Mix-Net contact platform and threshold security standard to preserve the elector identity. As e-voting advanced through years fraud activities like bribery, danger of buying votes, etc. increased. Hence new restriction features have been made such as coercion-resistance and receipt-freedom to address these issues. Coming to the improvements in recent years we have found a voting mechanism in Ref. [5] that depends on the email address of the voter which can be at a risk of Exploitation and Hacking. None of the important credentials such as protection and data security is assured by this approach. There is a high possibility of theft or modification of votes in this system proposed. A peer-to-peer Blockchain scheme was suggested in Ref. [6] by using a special model for vote commitment to Blockchain in order to preserve the privacy of the votes. According to another paper in Ref. [7], a system was developed in which blocks were generated after obtaining the votes from voters and maintained them in a database till the voting process ends. As per our knowledge and thinking, one of the good ideas proposed is explained by the article in Ref. [8] that describes the Hybrid method in which multiple chains have been incorporated in various levels and layers. It also focuses on the connections of the Internet of Things with the Blockchain technology.

In Ref. [9], a One Time Ring Signature was recommended to sustain the confidentiality of the vote, in this design, a public key was assigned to every voter leading to more power consumption by the CPU whenever a new voter is added, which made the signing process more complicated. A major drawback of this system is that we think it does not rely on any trusted base, but a better alternative would be making the government as the trusted base for the electoral process and giving them absolute powers for choosing the applicable politician. Biswas et al. [10] presented a unique lightweight proof of block and trade consensus method and integration framework for IoT blockchain. With this method, transactions and blocks may be validated in a shorter amount of time. In addition, the proposed framework offers a ledger distribution technique that may be used to reduce the memory needs of IoT nodes. The investigation and assessment of security factors, computing time, memory, and bandwidth requirements reveal a considerable increase in overall system performance.

A lightweight vehicular blockchain-enabled safe data exchange system in Unmanned Air Vehicle (UAV)-aided Internet of Vehicles (IoV) for disaster rescue has been proposed by Su et al. [11]. They attempted to accomplish three objectives in this paper. First, in disaster situations, they suggested a revolutionary UAV and blockchain-assisted collaborative aerial-ground network architecture. Second, in the lightweight vehicular blockchain, they build a credit-based consensus method to trace node behaviors and record data transactions safely and immutably with enhanced efficiency and security in attaining consensus. Third, they used trial and error to construct reinforcement learning-based algorithms to best optimize the pricing and quality of data sharing methods for both data contributor and data consumer.

Kim et al. [12] introduced a storage compression consensus (SCC) technique that compresses a blockchain in each device in order to assure storage capacity. When a lightweight device lacks sufficient storage capacity, it processes the SCC to compress the blockchain. They ran simulations to compare the SCC to the present consensus algorithm. As a consequence, as compared to the previous method, the suggested SCC lowered the storage capacity of the blockchain by 63%. By utilizing the blockchain approach, Yuan and Njilla [13] suggested a dependable decentralized reward system for crowdsensing. Unlike existing blockchain-based crowdsensing solutions that use expensive consensus mechanisms in terms of computation or financial cost, they investigated the power of the reputation system that exists in most crowdsensing applications and securely integrated it into blockchain to design a proof of reputation consensus mechanism. On top of that, an efficient and dependable incentive system based on blockchain technology was created. They used numerical analysis and simulation to evaluate the performance of the suggested incentive scheme.

Kaur et al. [14] proposed an effective cross-datacenter authentication and key-exchange scheme based on blockchain and elliptic curve cryptography. They employed distributed ledger of blockchain for maintaining the network information while the highly secure elliptic curve cryptography was used for mutual authentication between vehicles and road side units. The performance evaluation results against the existing state of the art revealed that the proposed scheme accomplished enhanced security features with reduced computational and communicational overheads.

10.3 Data Resources

JavaScript is used for the base coding of Blockchain. In our example, we used MetaMask [15] to Host our local Blockchain thorough Ethereum network. We used Node Package Manager for the truffle package in our project; it's used to create and manage files for creating smart contacts. We created two smart contracts with base language for the voting transaction in the Ethereum network. Ganache is used for simulating the local Blockchain network to create the addresses and private keys for the interaction with the Blockchain for the Ethereum network. HyperText Markup Language (HTML) and Cascading Style Sheets (CSS)are being used for the interaction between the decentralized network and the user.

10.4 Methodology

Blockchain is a decentralized ledger which can be viewed as a block of chains where each block contains a set of data. The process of adding a new block to the Blockchain is known as mining. The miners use Proof of Work (POW) algorithms to add new blocks to Blockchain. The blockchain for e-voting is shown in Figure 10.1. Each block can be identified by using a unique cryptographic hash. The block thus formed will contain a hash of the previous block, so that blocks can form a chain from the first block to the formed block linked with the help of linked list data structure. The voting block is shown in Figure 10.2. We created decentralized application using a public Blockchain where its transactions don't require any permission.

We created a decentralized network (peer-to-peer technology) because the data is not dependent on a single server point that holds the copy of all network configurations rather it passes the information through different node points which is difficult to access making it highly secured and providing privacy.

It is commonly referred as "trustless environment." It is easier to scale as we can add more computing power by simply adding more machine into this network. The decentralized network is shown in Figure 10.3.

The frontend of our website is developed using HTML/CSS and JavaScript which provides a user interface to access Blockchain data and give some inputs to Blockchain. It consists of one HTML/CSS page, one JavaScript page, and the interface which is provided by Google Chrome and MetaMask plug-in. The backend consists of smart contracts coded in solidity and a modifiable Ethereum Blockchain which is provided by Ganache. Two keys will be generated after the user enters his details that are public key and a private key. Public key is used for encryption whereas the private key is used for decryption. The person who receives the message can only see because he is having the private key that he can use to decrypt the file he received. Ganache is a personal Ethereum Blockchain that can be used to issue and execute commands, testing, and to perform transactions in Blockchain. The overview of directory is shown in Figure 10.4.

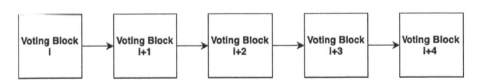

FIGURE 10.1
The Blockchain for e-voting.

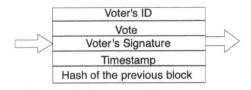

FIGURE 10.2
Voting block.

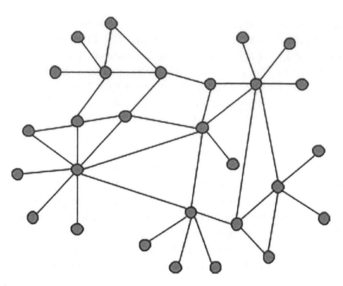

FIGURE 10.3
Decentralized network [16].

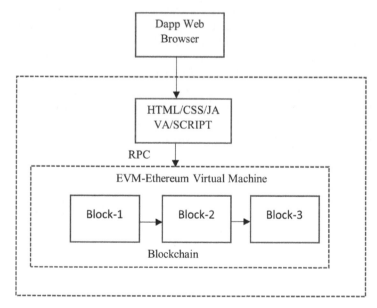

FIGURE 10.4
Overview of directory.

We used MetaMask which is a plug-in available in Chrome, used to build an Ethereum browser for managing decentralized applications and smart contracts. The image of MetaMask is shown in Figure 10.5, and the transactional address of Ethereum tool is shown in Figure 10.6.

Welcome to MetaMask

Connecting you to Ethereum and the Decentralized Web.
We're happy to see you.

FIGURE 10.5
Ethereum launching tool [17].

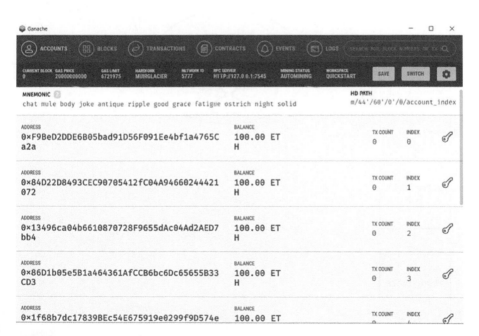

FIGURE 10.6
Transactional addresses of Ethereum network.

Index.html is a web page used to interact with Blockchain data. It is also used to display the number of votes along with the voter's information. The calculations in voting like the number of votes casted by individuals are performed by a JavaScript file App.js. The complete methodology of how our E-Voting system works is explained below with the help of Figure 10.7.

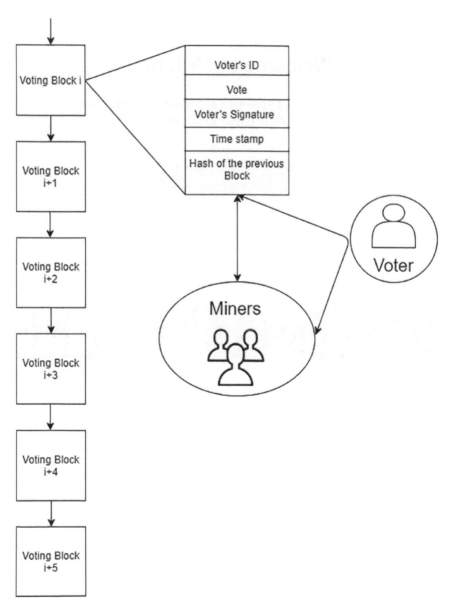

FIGURE 10.7
Voting process at a glance.

The user will choose the candidate they wish to vote for, by clicking on the vote button, where the transaction will be agreed automatically by MetaMask prompt. Then the user needs to click on submit for accepting the transaction which adds a new transaction in the Blockchain and disabling to vote one more time. Then the election results will be displayed.

10.5 Results and Discussion

We use Linux as a platform to implement and scheme and use JavaScript as the programming language for code. In each block there contains voter's id, vote, timestamp, and the hash of the previous block. As we know that people who are using phones and internet is very high we can assume that voter turnout will be high compared to the traditional voter system. And we use smart contracts which will be responsible for voters' procedures. It saves a lot of time and makes it easier to vote rather than the traditional way. We can register the accounts to vote and can restrict the unregistered accounts to vote. Since this is electronic voting system, we can eliminate the postal and some other delays which are caused by nature. In order to avoid the voter fraud, it maintains a very high-level security. Our process is easy: the voter will be given an address id and private key, and then you can log in and cast your vote which allows you to know quick updates about the elections. We made it more open, transparent, and secure. As we all know these days most of the young people do not vote even though they have the right to vote because it takes a lot of time to vote through postal ballot vote so though the electronic voting, we can assume that many people may vote since it's easy and secured.

10.6 Conclusion

The transparency that comes from decentralized networks using Blockchain helps to strengthen the voters' trust and reduce the election cost. It promises democracy and auditability. The idea of our project consists of a shared database network or distributed network that involves two completely separate Blockchains for voting data and electoral data on who voted. It also prevents the danger of aligning the votes from electors to specific candidates and also supports a tracking system to trace the elector details and obtain the correct number of votes. This system identifies that each vote is distinct with the help of ledger constituting the data of registered voters. After registration of the elector, the details would be legally verified or authenticated and then the vote will be assigned to him/her. Also a dual-check facility has been embedded so that no voter is compelled to vote. To eliminate unintentional votes the users are asked to verify their vote twice before the final submission. Without the access of the entire network infrastructure no intruder can obtain access to the votes collected. This feature is achieved through a data encryption system. The concept of secret keys helps in confirming the obtained electoral results that enables the Blockchain of votes to be reviewed by anyone and can be decoded with the latest added secret keys. Hence our efforts in this paper are to introduce a more efficient e-voting platform based on Blockchain technology where the results of the elections, privacy of the voter can never be manipulated and exploited. We aim for a fair, systematic, and transparent election process by using the Ethereum network to store data and to make the counting process in populated countries like India much faster by Parallel Computation.

Currently, the application only runs on the Ethereum network with peer-to-peer connections, which could be unreliable when implementing in real life. Because there is still a probability for any malpractice if some portion of the network might get collapsed as it's relying on unassured peer-to-peer networks connecting with the miners. So, in the future developmental process, the current decentralized network will be converted to distributed network, where the system doesn't need to rely on miners.

Acknowledgments

Our work could be achieved only with the encouragement and support from our precious institution Bennett University and our professor Dr. Vijay Bohat and HOD for their assistance and motivation. We are also thankful and respect the work contributed by many researchers as mentioned in Section 10.2 that laid a strong foundation for our project. We would really like to express our deepest gratitude to our parents, families, society, colleagues and our institution faculty members who offered us guidance and encouragement.

References

1. Chaum, D. L. Untraceable electronic mail, return addresses, and digital pseudonyms. *Communications of the ACM 24*, 2 (1981) 84–90.
2. DeMillo, R. A., Lynch, N. A., and Merritt, M. J. Cryptographic protocols. *In Proceedings of the Fourteenth Annual ACM Symposium on Theory of Computing* (1982), (pp. 383–400), ACM, San Francisco, CA, .

3. Fujioka, A., Okamoto, T., and Ohta, K. A practical secret voting scheme for large scale elections. *In International Workshop on the Theory and Application of Cryptographic Techniques* (1992), Springer, pp. 244–251.

4. Ohkubo, M., Miura, F., Abe, M., Fujioka, A., and Okamoto, T. An improvement on a practical secret voting scheme. *Information Security* (1999) (pp. 225–234). Springer, Berlin, Heidelberg.

5. Pawlak, M., Poniszewska-Marańda, A. and Kryvinska, N. Towards the intelligent agents for blockchain e-voting system. *Procedia Computer Science*, 141 (2018) 239–246.

6. Tarasov, P. and Tewari, H. The future of e-voting. *IADIS International Journal on Computer Science and Information Systems*, 12(2) (2017) 148–165.

7. Hanifatunnisa, R. and Rahardjo, B. Blockchain based e-voting recording system design. *2017 11th International Conference on Telecommunication Systems Services and Applications (TSSA)*, Lombok (2017) pp. 1–6.

8. Bartolucci, S., Bernat, P. and Joseph, D. SHARVOT: Secret SHARe-based voting on the Blockchain. *2018 IEEE/ACM 1st International Workshop on Emerging Trends in Software Engineering for Blockchain (WETSEB)*, (2018) pp. 30–34.

9. Reyna, A., Martín, C., Chen, J., Soler, E. and Díaz, M. On blockchain and its integration with IoT. Challenges and opportunities. *Future Generation Computer Systems*, 88 (2018) 173–190.

10. Biswas, S., Sharif, K., Li, F., Maharjan, S., Mohanty, S.P. and Wang, Y. PoBT: A lightweight consensus algorithm for scalable IoT business blockchain. *IEEE Internet of Things Journal*, 7(3) (2019) 2343–2355.

11. Su, Z., Wang, Y., Xu, Q. and Zhang, N. LVBS: Lightweight vehicular blockchain for secure data sharing in disaster rescue. *IEEE Transactions on Dependable and Secure Computing*, 19 (2020), 19–32.

12. Kim, T., Noh, J. and Cho, S. SCC: Storage compression consensus for blockchain in lightweight IoT network. *In 2019 IEEE International Conference on Consumer Electronics (ICCE)* (2019) pp. 1–4. IEEE.

13. Yuan, J. and Njilla, L. Lightweight and reliable decentralized reward system using blockchain. *In IEEE INFOCOM 2021-IEEE Conference on Computer Communications Workshops (INFOCOM WKSHPS)*, (2021) pp. 1–6. IEEE.

14. Kaur, K., Garg, S., Kaddoum, G., Gagnon, F. and Ahmed, S. H. Blockchain-based lightweight authentication mechanism for vehicular fog infrastructure. *In 2019 IEEE International Conference on Communications Workshops (ICC Workshops)*, (2019) pp. 1–6. IEEE.

15. https://MetaMask.io// [Date of Access-12–12–2021].

16. Truffle Suite. "Ganache: Overview: Documentation." Truffle Suite, www.truffle suite.com/docs/ganache/overview [Date of Access-11–12–2021].

17. MetaMaskDocs "#EthereumProviderAPI." docs.MetaMask.io/guide/ethereum-provider.html [Date of Access-15-12-2021].MetaMask

Index

Note: **Bold** page numbers refer to tables and *italic* page numbers refer to figures.

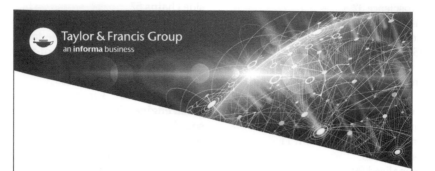